# COMMENTARY ON THE 15th EDITION OF THE IEE WIRING REGULATIONS

B. D. Jenkins

**Co-operating organisations**

The Commentary has been produced with the co-operation and support of The Electrical Contractors' Association (ECA), the Electrical Contractors' Association of Scotland (ECAS), The Electricity Council (EC), The Institution of Electrical and Electronics Technician Engineers (IEETE), The Institution of Electrical Engineers (IEE) and The National Inspection Council for Electrical Installation Contracting (NICEIC). The author and publisher are grateful to these organisations and to their members who formed the following Editorial Advisory Panel:

D C Brice (IEE)
W Cartwright (EC)
T Hedgeland (NICEIC)
T Marshall (ECAS)
P S Pontin (IEETE)
J F Wilson (ECA)

Their advice and assistance have been invaluable.

# COMMENTARY ON THE 15th EDITION OF THE IEE WIRING REGULATIONS

B. D. Jenkins

*with the co-operation and support of*

The Electrical Contractors' Association
The Electrical Contractors' Association of Scotland
The Electricity Council
The Institution of Electrical and Electronics Technician Engineers
The Institution of Electrical Engineers
The National Inspection Council for Electrical Installation Contracting

**PETER PEREGRINUS LTD.**
on behalf of the
Institution of Electrical Engineers

Published by: Peter Peregrinus Ltd., Stevenage, UK, and New York



**British Library Cataloguing in Publication Data**

Jenkins, B.D.
A commentary on the 15th edition of the IEE Wiring Regulations.
1. Electric wiring
I. Title
621.319'24 TK 3201

**ISBN 0-906048-51-6**

Printed in the United Kingdom by A.McLay & Co. Ltd.,
Cardiff and London

# Contents

# Preface

Having been closely concerned over the past six years with the preparation of the Fifteenth Edition of the IEE Regulations for Electrical Installations, I accepted the invitation to write this commentary on the new edition with somewhat mixed feelings. However, a number of organisations which are closely in touch with the work of the IEE Wiring Regulations Committee expressed their willingness to co-operate in the preparation of this Commentary and I have been fortunate in having the assistance of a small Editorial Advisory Panel of nominees of the co-operating organisations listed on the title page.

I gratefully acknowledge the extensive help and advice I received from the members of that Panel, and I am sure the other members will not object if I express particular thanks to Mr. W. Cartwright of the Electricity Council and Mr. T. Hedgeland of the National Inspection Council for Electrical Installation Contracting. I also wish to acknowledge the considerable help and encouragement I received from my colleagues at the Institution of Electrical Engineers and the permission I was readily given to use extensive notes developed during the preparation of the Fifteenth Edition itself. My thanks also go to others from whom I sought advice and which was, without exception, always freely given and any success enjoyed by this Commentary will be due in no small measure to all those who helped me in my task.

In writing this Commentary I have attempted, and I hope succeeded, in producing a readable introduction to the Wiring Regulations and, at the same time, a book which subsequently is equally useful as a reference book. It has been aimed primarily at installation designers, inspectors and supervisors but it is hoped that anyone concerned in electrical installation work, whether on the design side or in the on-site erecting and installing, will also find something of interest in this Commentary.

Every care has been taken to ensure the accuracy of this Commentary but I must emphasise that the explanations and views expressed are my own and do not purport to represent the official views of either the Institution of Electrical Engineers or of the co-operating organisations; in particular, they are without prejudice to any official interpretations of the Wiring Regulations that may be given by the IEE Wiring Regulations Committee.

Brian D. Jenkins

Twickenham

# Introduction

The Society of Telegraph Engineers and of Electricians published in 1882 their 'Rules and regulations for the prevention of fire risks arising from electric lighting' and these are recognised as the first edition of the IEE Wiring Regulations. This first edition, together with a foreword and details of the membership of the drafting committee, took up a mere four small pages of print. Some readers, it is felt, will be interested to see this first edition and for this reason it has been reproduced as Appendix A of this Commentary.

But in the intervening near century, electrotechnology has been in a state of continual and sometimes rapid evolution. In order to keep pace with this, the Regulations have been kept under constant review and have been periodically revised, this work being undertaken by the IEE Wiring Regulations Committee on behalf of the Council of the Institution.

The Wiring Regulations Committee are a small number of members of the Institution who serve as individual experts together with representatives of interested organisations such as other professional institutions, government departments, nationalised industries and electrical contractors' and manufacturers' associations.

Undoubtedly the approach and style used in this latest edition will be very strange to those concerned with electrical installation work because both are so different from previous editions, but it is believed that as designers and installation contractors become more familiar with the basic plan by actually designing and erecting installations to the new edition any initial misgivings or objections they may have will largely disappear.

What is important to the designer and contractor is that not only have present practices been recognised in the Fifteenth Edition but, in addition, a wider range of options is now available to them, and they will find a greater degree of freedom than before to devise their own means of complying with requirements which are explicitly and quantitatively prescribed.

However, this greater freedom is accompanied by greater responsibility and the designer who wishes to gain the maximum advantage from the Fifteenth Edition will require more data on the components of his intended installation than has been needed when designing to previous editions. In the Fifteenth Edition, considerable use has been made of appendixes in order to provide some of that data, and standard circuit arrangements have also been included which can be used, particularly in the smaller installation such as the typical household installation, where a detailed design may be considered unjustified.

The use of standard circuit arrangements, as will be seen later, does not absolve the designer or contractor from having to carry out some technical assessment and calculation but, in order to assist users of the IEE Wiring Regulations, one chapter of this Commentary is devoted to an examination of the implementation of the Fifteenth Edition in simple installations and the impact of this edition on well established practices in such installations.

That chapter also takes the reader through the electrical design of a final circuit of the commonly encountered type where the protective device (fuse or miniature circuit breaker) is intended to give protection not only against overload and short circuit currents but also protection against electric shock in case of a fault (now termed protection against indirect contact).

Earlier in this Introduction, 'the designer' has been mentioned either solely or coupled with 'the contractor'. While it is recognised that there are many small companies in the electrical contracting industry where the owner does the design as well as the actual erection work on site and the amount of designing he does may be minimal, it must not be forgotten that, in the end, it is his responsibility to ensure that any claim that the installation concerned complies with the Wiring Regulations is justified and can, if so demanded, be substantiated.

The remainder of this Commentary examines numerous individual regulations, where it is believed that some explanation or further information is necessary, or is at least desirable, for a fuller understanding by the reader of the intention behind those particular regulations.

Although the Fifteenth Edition unquestionably exhibits a considerable departure from preceding editions in both approach and style, its real importance stems from the fact that it is the first to be mainly based on *internationally* agreed installation rules, the two international bodies concerned being the world-wide International Electrotechnical Commission (IEC) and the regional European Committee for Electrotechnical Standardization (CENELEC).

The IEC was formed in 1906 with a membership of fourteen countries, or to be more precise, of the specially formed 'National Committees' of those countries, and in the intervening years the membership has steadily grown to over three times the original number of National Committees.

CENELEC was formed in 1973 when the United Kingdom, in company with Denmark and Eire, joined the European Economic Community and now comprises the National Committees of the countries of the enlarged Community together with those of the European Free Trade Association and Spain.

For the United Kingdom, the National Committee for both international standards bodies is the British Electrotechnical Committee of the British Standards

Institution and the IEE Wiring Regulations Committee acts on behalf of, and through, the British Electrotechnical Committee in the work of both the IEC and CENELEC technical committees dealing with electrical installation rules. Fig. 1 shows the close relationship which exists between the British Standards Institution and one of its 'Founder Institutions'–The Institution of Electrical Engineers.

For the purpose of this Commentary on the IEE Wiring Regulations it is not necessary to describe the detailed organisation of the two international standards bodies or the procedures they use for the preparation of standards, but it is

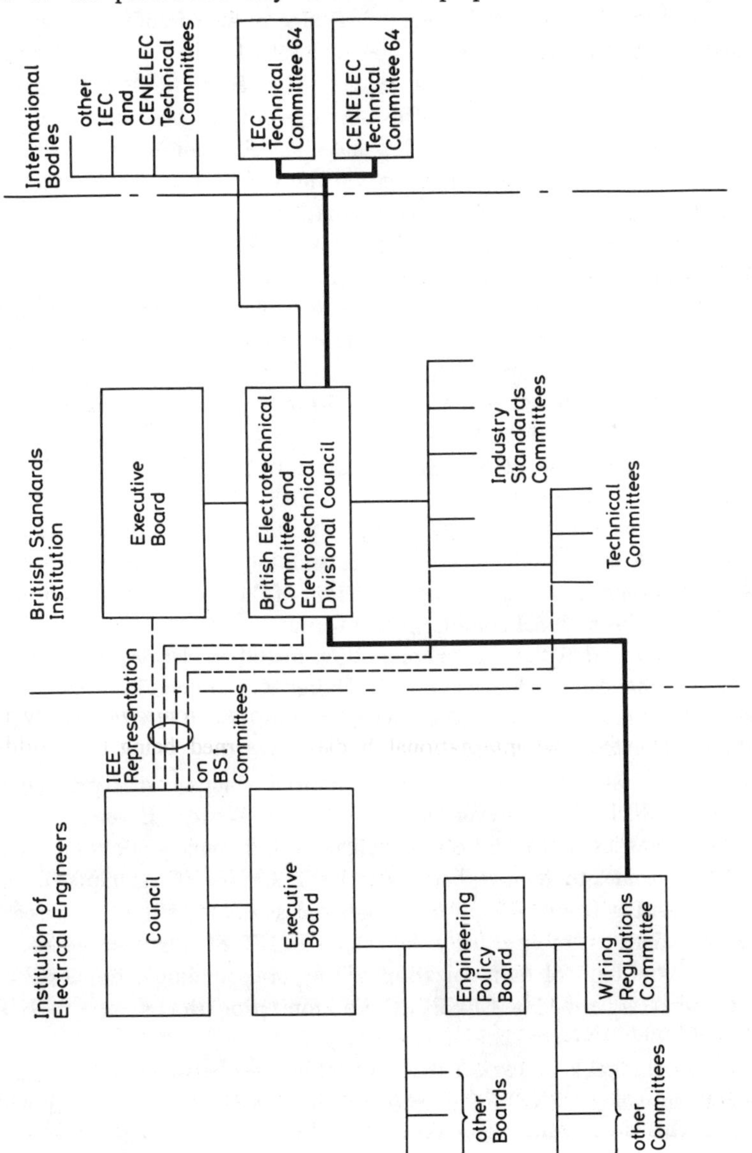


**Fig. 1** ***Relationship between the IEE, the BSI and international standards bodies***

important for the reader to be aware of certain aspects of the inter-relationship between the international standards and national standards (including the national wiring rules which, for the United Kingdom, are the IEE Wiring Regulations) and between these standards and legislation, both national and international.

IEC set up a technical committee (TC 64) in 1968 to commence work on formulating international rules for the electrical installations of buildings and it was soon realised that agreement would not be reached by attempting to combine the existing national requirements because, although the general standards of safety were broadly the same among the technologically advanced countries having well established national wiring regulations and a few basic rules seemed to be superficially similar, these regulations had tended to develop into sets of apparently arbitrary standardised arrangements. This development had been influenced by the legal, social and commercial requirements and climatic conditions of the individual countries leading to significant differences in the various national wiring rules, so that, when attempting to formulate the international rules, it proved impossible to consider any one particular aspect without taking account of its relationship to others and of the particular historical background concerned.

IEC Technical Committee 64 therefore decided that in order to establish internationally acceptable wiring rules it would be necessary to go back to basic fundamentals and this objective approach led almost inevitably to the plan and format adopted for the international standard and followed by the IEE Wiring Regulations Committee for the Fifteenth Edition.

The status and purpose of IEC Standards are stated in the foreword of every such standard:

> 'An IEC Standard expresses, as nearly as possible, an international consensus of opinion on the subject dealt with and the wish is expressed that all National Committees should adopt the text of the Standard for their national rules insofar as national conditions will permit.'

It is seen, therefore, that there is no compulsion placed on National Committees to adopt any IEC Standard and, in fact, in the United Kingdom, as in other countries, it has sometimes been necessary to introduce possibly significant departures from an IEC Standard in the corresponding national standard.

While the IEC standards work is very important in its own right, there is a further aspect which makes it even more so for most Western European countries because the countries of the European Economic Community are committed, by the Treaty of Rome, to remove barriers to trade. CENELEC attempts to remove such barriers arising from differences in national standards either by preparing a European Standard or, more usually, by taking an IEC Standard as the 'reference document' for the basis of harmonization of the corresponding national standards, leading to what is termed a CENELEC Harmonization Document (or 'HD' for short).

It is not unreasonable to expect that any differences between an IEC Standard and the corresponding CENELEC Harmonization Document would be minimal but if such differences (termed 'common modifications') do occur, then by the

CENELEC rules the HD must clearly indicate what they are. However, in order to reach agreement in a harmonization project, CENELEC has accepted that sometimes there may have to be not only common modifications to the reference document but also what are termed 'national deviations'. These deviations may be caused by national legislation, regulations or administrative rules or by network conditions (such as supply voltage or system earthing), climatic conditions and the like. Such deviations are called 'A-deviations' and are recognised as being outside the competence of the National Committees.

There is a second type of national deviation and this is a specific technical requirement in a national standard which differs from, or is additional to, the corresponding Harmonization Document, but which the responsible National Committee considers necessary to retain for the time being. Such a deviation is termed a 'B-deviation' and is only permitted if considered justifiable by the other National Committees.

In all CENELEC harmonization work, national deviations, whether Type A or Type B, are intended to be retained for only a limited period, although in the case of the former the National Committees can only undertake to endeavour to persuade the competent authorities in their countries to change the relevant national laws to enable these Type A deviations to be discarded.

Under CENELEC rules each National Committee formally undertakes to take the following action when a Harmonization Document has been issued:

(*a*) to issue a national standard covering the same subject, and complying with the Harmonization Document, or

(*b*) to withdraw any national standard by an agreed date if it conflicts with the Harmonization Document.

So here too there is no compulsion on National Committees to adopt a Harmonization Document, but if there is a corresponding national standard it must not be in conflict with that Harmonization Document.

IEC Technical Committee 64, using the objective approach referred to earlier have reached agreement on a number of chapters and parts of IEC Publication 364 'Electrical installations of buildings' and are at an advanced stage of discussion on others. The chapters and parts already agreed are being used, or have been used, as the reference documents for corresponding Harmonization Documents. When agreement is reached in CENELEC the national wiring rules are required to be brought into alignment with the relevant Harmonization Documents and this has been done in the Fifteenth Edition, as indicated in its Preface.

It must be pointed out that, although the work in IEC and CENELEC is far from completed, it was decided to adopt the IEC plan and the technical content of those chapters already agreed in IEC, where necessary in advance of harmonization in CENELEC, in the expectation that they would be accepted by the latter body with the minimum number of common modifications.

For those chapters or sections on which no firm international agreement has yet been reached, it has been necessary to retain the existing national regulations, i.e. those of the Fourteenth Edition. This has been done in order to respect an agreement in CENELEC to avoid as far as possible the unilateral introduction of new national requirements which might be prejudicial to eventual international agreement on the particular aspects concerned.

Finally, in this brief Introduction it is appropriate to explain why it was considered necessary to publish a new edition of the Wiring Regulations at this time. Some of the concepts developed in the process of formulating the international wiring rules have also been used in the preparation of the international safety standards for electrical equipment. Those concepts will therefore be absorbed into the British Standards corresponding to those international equipment standards and it would be obviously ridiculous if the IEE Wiring Regulations did not include the same concepts as this could well lead to unacceptable incompatibility between the equipment requirements and those regulations.

Then again, British installation contractors, consulting engineers and equipment manufacturers involved in overseas projects in countries adopting the international requirements for installations must surely be helped if the British wiring rules also followed the same requirements. In any event, some countries have traditionally been influenced by British electrical installation practice and such countries must view the British implementation of these international requirements with considerable interest. If that implementation was not as complete as possible there could be a lessening of British influence in the future.

A criticism of previous editions of the Wiring Regulations was that they were heavily slanted to installations in domestic dwellings and similar relatively simple installations but did not cover the larger, more complex, ones. The Fifteenth Edition, it is believed, goes at least some way to countering that criticism although it is recognised that the regulations will still have to be augmented for heavy industrial installations, for example, for the process industries and for other specialised installations.

Some might argue, while accepting the need to align with internationally agreed rules, that one should have waited until the work on those rules had reached a more advanced stage. The answer to that argument is readily given. It must be remembered that the Fourteenth Edition of the IEE Wiring Regulations was published in 1966 and although amendments to that edition were issued in 1970, 1974, and 1976, fifteen years is probably as long a life as one could reasonably expect from any edition, particularly when in that period there have also been significant changes in associated equipment standards and in installation practices.

The reader of this Commentary will find that relatively few comparisons have been given between the new edition and the Fourteenth Edition. This has been done deliberately because it is believed that if such a method had been more widely adopted it would have only delayed the process of assimilation of the new approach embodied in the Fifteenth Edition. In any event, the newcomer to electrical installation practice does not need such retrospective treatment – he is concerned with the present and future requirements, not with the past – and it is hoped that this Commentary will be of particular assistance to those who are at the beginning of their careers in this important field of electrical engineering.

Chapter 1

# Plan, style and terminology of the Fifteenth Edition

As indicated in the Introduction, IEC Technical Committee 64 found it necessary to go back to basic fundamentals in order to develop international rules and this led them almost inevitably to the plan which has also been adopted for the Fifteenth Edition.

## 1.1 Plan and style of the Fifteenth Edition

Fig. 2 shows diagrammatically the correlation between the various Parts, Chapters and Sections of the Fifteenth Edition. Of course, some other plan could have resulted, but a particular advantage of the plan which has been used is that it is a logical one although it is recognised that some users of the Fifteenth Edition may initially find it strange and possibly difficult to follow.

This difficulty, it is believed, will be largely caused by the fact that, although the plan is logical in the sequence adopted for the consideration of the various risks leading to the prescription of the acceptable methods of protecting against those risks and then of the devices one is permitted to use, the plan does not necessarily follow the sequence of design steps used by the installation designer.

For instance, although the first step in designing an installation must always be to carry out the assessment of characteristics demanded by Part 3 of the Wiring Regulations, including the determination of the number of circuits to be used for compliance with Section 314, the designer may then decide to adopt the following sequence of design steps:

(*a*) Choose the types of cable which the Wiring Regulations permit him to use, taking account of the requirements prescribed in Section 523 appropriate to the environmental conditions expected.

(*b*) Determine the cross-sectional areas of the conductors of those cables to carry the expected load currents for compliance with *Regulation 522-1*, taking into account the method of installation of the cables, grouping with other cables, and so on.

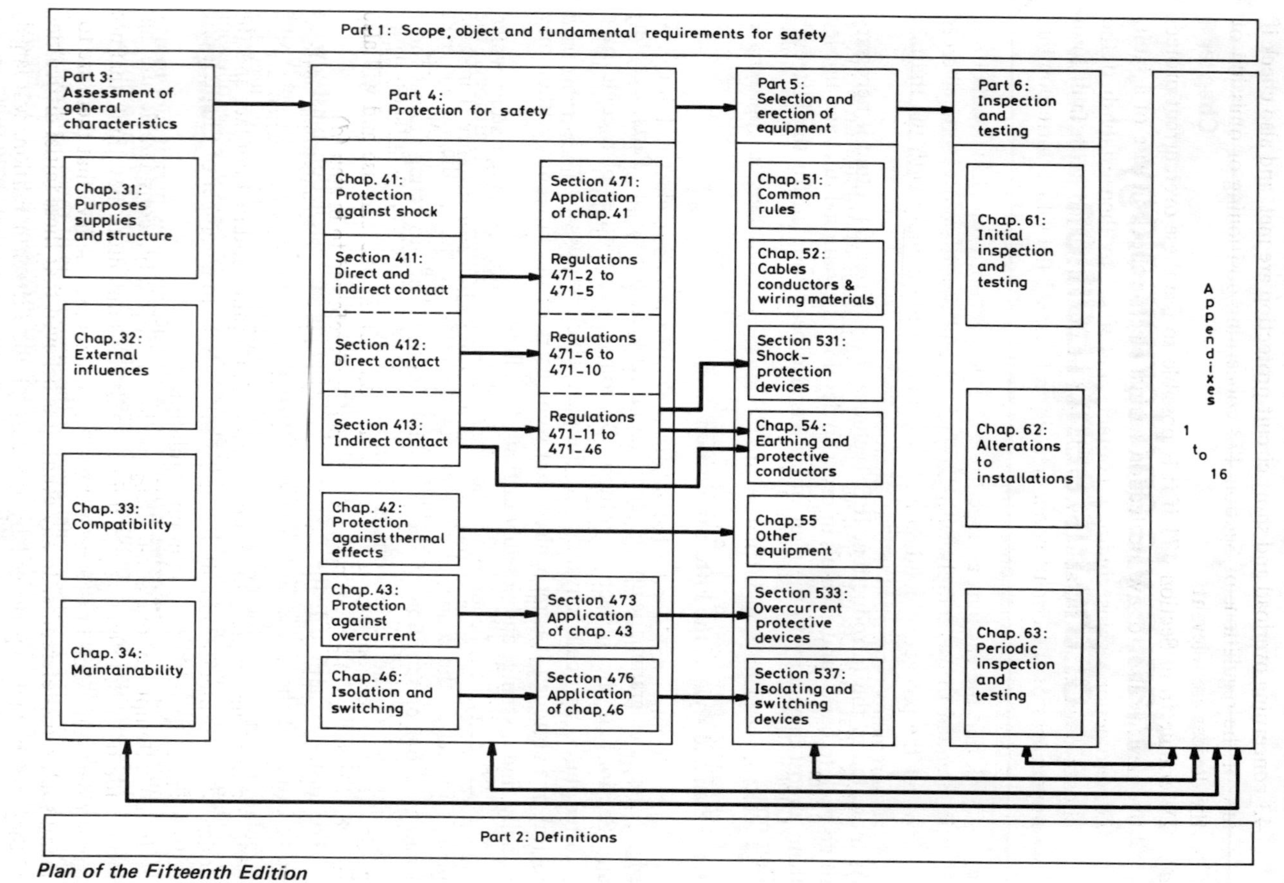


**Fig. 2** *Plan of the Fifteenth Edition*

(*c*) Choose the types of overcurrent protective device to use and determine their rated currents. Then establish whether the requirements prescribed in Chapter 43 concerning overload and short circuit protection are met, and also check if any of the regulations of Section 473 concerning positioning or omission of the devices are relevant.

(*d*) Determine from Section 471 if it is possible to use these overcurrent protective devices also to give protection against electric shock in case of a fault, i.e. protection against indirect contact, and if so, to then establish if the chosen devices are suitable by checking that the calculated earth fault loop impedances of the circuit concerned do not exceed the values corresponding to the relevant disconnection times prescribed in Chapter 41.

(*e*) Check that the earthing arrangements and the types of protective conductor it is intended to use comply with Chapter 54 and that the cross-sectional areas of all the protective conductors in the installation comply with the thermal requirements prescribed in that Chapter.

(*f*) Check that the circuits comply with the limitation of voltage drop under normal load conditions as prescribed in *Regulation 522-8*.

(*g*) Choose the means of isolation it is intended to use, the devices for switching off for mechanical maintenance and for emergency switching, as appropriate, as required by Chapter 46 and Section 476.

Of course, having established what might be termed the basic electrical design of the installation, it is then necessary to check the other aspects of the proposed design to verify that they also comply with the relevant regulations, in particular the measures taken for protection against direct contact and the precautions used to prevent mechanical damage.

As already stated, the foregoing is only one sequence of design steps which may be used because other sequences are equally possible. For instance it may be argued that step (*f*) should be combined with, or immediately follow, step (*b*) because the limitation of voltage drop may well be the determining factor as regards the cross-sectional areas of the cable conductors one can use and will also therefore influence the earth fault loop impedances referred to in step (*d*).

Another possible cause of difficulty in using the Wiring Regulations, again at least initially, is the fact, as indicated in steps (*d*) and (*e*), that even when dealing with one particular aspect, in this case protection against indirect contact and the design of the protective circuit, it is necessary to refer to a number of Chapters and Sections in order to find the relevant regulations.

In an attempt to overcome that particular difficulty, considerable use has been made of cross-referencing in the new edition; some users may think that not enough has been done in this respect whereas others may believe that there are too many such cross-references. In any event, it is believed that these initial difficulties, if they are in fact experienced by the user of the Fifteenth Edition, will largely disappear as one becomes more familiar with it and begins to appreciate that it really is a logical and objective approach to the problem of presenting reasonably concise rules intended to cover a wide range of different types of electrical installation.

It will help the user of the Fifteenth Edition if he remembers that, in outline, the plan is as follows:

Part 3 deals with the identification of the characteristics of the installation, of its environment and of the source of energy supplying the installation.

Part 4 describes in Chapters 41 to 46 the basic measures of protection against the various risks and in Chapter 47 gives the circumstances for which particular measures are suitable.

Part 5 deals with the requirements for the selection and erection of equipment so as to comply with Parts 3 and 4.

Part 6 covers the inspection and testing that is required to be carried out to check that the installation concerned complies with the Wiring Regulations.

Part 4 is probably going to be the least easy for the user of the Fifteenth Edition to become accustomed to, but even so, there should be little difficulty if it is remembered that, having first established that the Regulations allow the use of the protective measure it is intended to adopt, it is not then necessary to observe any requirements relating to the other protective measures which might have been used as an alternative. For example, protection against electric shock by the use of safety extra-low voltage (SELV) is a complete measure in itself, and for parts of an installation in which that measure is used no account need be taken of requirements in Sections 412 and 413 for protection against electric shock by means of other measures.

To take a more common example, the overcurrent protective device (e.g. a fuse or miniature circuit breaker) used to give protection to a final circuit against short circuit currents is very commonly used to also give protection against indirect contact, this being one example of the measure termed 'automatic disconnection of the supply'. In a particular case, having checked that the pertinent regulations (such as that concerning the maximum value of earth fault loop impedance which can be tolerated) are met, there is no need to consider the regulations in Section 413 covering the other measures for protection against indirect contact.

Although, in general, this Commentary considers the regulations in the same order in which they appear, for commenting on Part 4 of the Wiring Regulations the approach adopted has been to deal with the four main aspects of that Part as separate themes, each theme being self-contained as a Chapter of this Commentary. For example, in considering protection against electric shock the Commentary considers Chapter 41, Section 471 and Chapter 54 of the Wiring Regulations together and not separately.

## 1.2 Definitions in the Fifteenth Edition

An integral and essential part of any set of rules is that which gives the meanings of the terms used, namely the definitions.

In the development of the international wiring rules, it was found necessary also to develop a new terminology and the adoption of those rules into the Fifteenth Edition meant introducing the new terms. Thus in this edition the user will find a number of terms which will be strange to him but to which he must become accustomed, as a knowledge of their meanings is essential to a full understanding of the regulations in which those terms are used. At the same time he will fail to find some of the terms used in previous editions; to quote one example, the earth continuity conductor (e.c.c.) of a twin-core and earth cable is now a 'circuit protective conductor', which surely will soon be called a 'c.p.c.'.

Initially, the reader of the Fifteenth Edition may experience some difficulty in distinguishing between the various types of 'protective conductor', this being the generic term which embraces, in addition to circuit protective conductor, 'combined neutral and protective conductors', 'main equipotential bonding conductors', 'supplementary bonding conductors', 'earth-free equipotential bonding conductors' and 'earthing conductors'. It is important that a protective conductor is correctly identified, because its function will determine which regulations, particularly those in Chapters 41 and 54, are applicable, but these protective conductors are considered in detail when the chapters of the Wiring Regulations are themselves considered.

Some terms previously used have disappeared altogether because the method of approach for a particular design aspect has changed completely. The best example of this is the loss of 'close' and 'coarse' excess-current protection, and these terms have not been replaced by new ones because the method for determining the cross-sectional areas of cable conductors is now somewhat different from that used in earlier editions.

Throughout this Commentary an explanation of some of the new terms is given, where considered appropriate, and therefore in this present chapter there is no need to make further comment on them. However, there are two terms which are of particular interest because they are used in the Fifteenth Edition in a different sense from that in previous editions.

The first of these is *'electrical installation'* which, unless qualified, can be applied to either a complete installation or any section of it. In respect of certain requirements of the Wiring Regulations, particularly those relating to isolation and switching and parts of the requirements for protection against electric shock, it may be necessary to consider the complete installation as a collection of separate installations each having its own origin. As a consequence, the term 'origin of an installation' is not necessarily confined to a single point at which energy is delivered from a public supply network.

It will be noted that the term used here is 'network' and not 'system', because in the Wiring Regulations the latter has a particular meaning, denoting as it does the combination of a 'source of energy' and an 'installation'. As indicated in Appendix 3 of the Wiring Regulations which gives explanatory notes on the various types of system earthing, the most commonly encountered system is where the 'source of energy' is the public supply network. Even in a large industrial complex when the source of energy, the wiring and current-using and other equipment are all owned or controlled by the user of the premises and constitute what is usually called an installation, it is preferable to consider this combination as a 'system' and the source of energy treated as being separate from the other equipment in the premises.

The second term is *'live part'*, which now includes a neutral conductor (other than a combined neutral and protective conductor). The term 'phase conductor' is used wherever it is intended to refer to the conductors(s) of a circuit excluding the neutral.

There are some new terms included in the Fifteenth Edition which have not emanated from the IEC work on international wiring rules but have been adopted from other sources. As just one example of this, the term 'current-operated earth-leakage circuit breaker' has been replaced by 'residual current circuit breaker' which is one type of device covered by the generic term 'residual current device' which embraces all devices which rely on the detection of the residual current flowing in a circuit in the event of an earth fault, residual current being defined as the r.m.s. value of the vector resultant of the instantaneous values of the currents flowing in the main circuit of the device.

Chapter 2

# Scope, object and fundamental requirements for safety

## 2.1 Chapter 11: Scope

The types of installation which were specifically in mind in drafting the Wiring Regulations are given in *Regulation 11-1*, although the Regulations are widely used for other applications in the absence of other suitable rules, one example being street lighting installations. Contracts for such installations usually require compliance with the requirements of the Wiring Regulations considered to be relevant, such as those concerning cable sizes, the provision of overcurrent protection and earthing.

Although the Institution welcomes this use of the Wiring Regulations, in all such cases it is the responsibility of the specifier (who should be a suitably qualified professional engineer) to judge the suitability of the Wiring Regulations and to authorise any modifications or additions considered to be necessary.

The ranges of voltages covered by the Wiring Regulations are identified in *Regulation 11-2*, the voltage bands defined there being those agreed internationally (and which are given in CENELEC HD 193). These voltage bands, however, were first introduced into the Wiring Regulations in the 1976 Amendments to the Fourteenth Edition and are therefore not new.

Although at the time of writing the statutory regulations referred to later, such as the Electricity (Factories Act) Special Regulations 1908 and 1944, still retain the voltage bands known as 'medium voltage' which covers voltages above 250 V but not exceeding 650 V and 'high voltage' for voltages exceeding 650 V but not exceeding 3000 V, it is probable that the revision of those regulations will adopt the internationally agreed voltage ranges. In considering the extra-low voltage range, the Wiring Regulations now differentiate between what is termed functional extra-low voltage and safety extra-low voltage, but these will be commented on later in some detail.

In previous editions, equipment operating at voltages not exceeding 15 V to earth or between phases was not generally required to comply with the Wiring

Regulations, provided that it was suitably segregated from other equipment working at higher voltages. In the Fifteenth Edition, that exclusion no longer exists and the scope of the Edition covers installations for all nominal voltages up to the limit of low voltage.

Where the Wiring Regulations are used for purposes other than those specifically mentioned in *Regulation 11-1,* special care should be taken if the nominal voltage exceeds the upper limit of low voltage. The only two applications where the nominal voltage exceeds low voltage which are currently covered by the Wiring Regulations are discharge lighting and electrode boilers.

Exclusions from the scope are listed in *Regulation 11-3* but only some of them require commenting on here. For instance, Item (i) indicates that the Wiring Regulations do not apply to public supply networks and the responsibility of the installation designer and contractor begins at the origin of the installation, generally the point at which the supply of energy is delivered from the public supply network. Nevertheless, as indicated in *Regulation 313-1* it is the responsibility of the designer and contractor to take account of those characteristics of the network which affect the design of the installation, two of these characteristics being the prospective short circuit current and the contribution of the public network to earth fault loop impedance.

Item (ii) indicates that the Wiring Regulations have only a limited application to installations in potentially explosive atmospheres, and it is essential that the designer also observes the recommendations in the British Standard Codes of Practice CP 1003 and BS 5345. The latter, which when complete will consist of twelve Parts, is intended to supersede CP 1003 and although Part 1 of BS 5345 which gives the basic requirements for all Parts of the Code does not at present make any reference to the IEE Wiring Regulations, some of the other Parts do.

Item (iii) indicates that the Fifteenth Edition, unlike previous editions, does apply to the telecommunication and other circuits to which reference is made except, as stated, those supplied by safety sources. Not all the requirements of the Wiring Regulations may be found applicable to such circuits but the circuits are notably expected to comply with the rules for protection against electric shock (for example the rules regarding supplies for, and segregation of, safety extra-low voltage circuits).

Fire detection and alarm systems in buildings are covered by the Code of Practice BS 5839 and emergency lighting of premises by BS 5266. Intruder alarms are the subject of another Code of Practice which is in preparation. While considerable care is taken to avoid conflict between the Wiring Regulations and these Codes of Practice it must be remembered that in buildings which are so-called ‘designated buildings’ under the Fire Precautions Act the prime requirement would be to effect compliance with the Codes of Practice. The Codes therefore take precedence over the IEE Wiring Regulations where there are minor differences. The segregation requirements referred to in Item (iii) are prescribed in *Regulations 525-1 to 525-9.*

While the Wiring Regulations, as indicated by *Regulation 11-4*, do not include constructional requirements for equipment and *Regulation 511-1* requires that all installation equipment shall comply with the appropriate British Standard, or as indicated later, with an equivalent foreign standard, some provisions are included for equipment which is constructed on site in the process of erecting an installation.

Factory-built assemblies (FBAs) are generally treated as complete items of equipment complying as a whole with BS 5486, but for various reasons such as transport difficulties or production methods certain steps of the assembly of an FBA may have to be made on site. When this happens, such assemblies are still considered to be factory built, provided that the contractor when completing the assembly on site does so strictly in accordance with the manufacturer's instructions in order that compliance with the British Standard is maintained.

The construction on site of certain items of equipment instead of using factory-built assemblies may result in the installer having to carry out tests on completion of those items which are broadly similar to those used in the manufacture of the latter assemblies but it may be found inconvenient or impracticable to make such tests on site. For instance, where protection against indirect contact is to be attained by the protective measure termed 'Protection by use of Class II equipment or by equivalent insulation', one acceptable method is to apply supplementary insulation to electrical equipment having basic insulation only, and another is to apply reinforced insulation to uninsulated live parts. Where either method is used, an applied high voltage test must be carried out similar to that required by the British Standard for similar factory-built equipment (see *Regulations 413-18, 413-22* and *613-10*) and because of this and other practical difficulties which may be encountered there is no doubt that the use of factory-built assemblies is to be preferred in most cases.

*Regulation 11-5* takes account of the types of temporary prefabricated installation such as festoon lighting which are intended to be used separately from one site to another or dismantled and re-erected seasonally. While such installations are required to meet the requirements of the Wiring Regulations concerning, for instance, protection against electric shock (Chapter 41), protection against thermal effects (Chapter 42), protection against overcurrent (Chapter 43) and isolation and switching (Chapter 46) and the equipment used must comply with the relevant regulations, *Regulation 11-5* recognises that in general the erection requirements of Chapter 52 are not applicable.

In designing these installations, it must be borne in mind that their erection is often undertaken by persons who have no electrical training and account must be taken of this fact. In many cases too, the manner in which these installations are used is far from satisfactory and the attention of the designer is drawn to the particular importance of *Regulation 341-1* which requires an assessment to be made as to the frequency and quality of maintenance that the installation can be expected to receive.

## 2.2 Chapter 12: Object and effects

The regulations which are in this Chapter appeared in previous editions in the Introduction to the Wiring Regulations and not in the Regulations themselves. It will be noted that, except for *Regulations 12-3, 12-5* and *12-7*, the regulations here are informative in nature. For instance, *Regulation 12-1* states that the basic object of the Wiring Regulations is 'to provide safety especially from fire, shock and burns' and comment is not really needed here, this always having been the object of the Regulations.

The Fifteenth Edition, like previous editions, does not refer to the functioning of the installation as such or to the efficiency of that functioning. However, a number of individual regulations have a direct bearing on this particular aspect, one example being *Regulation 522-8*, which limits the voltage drop under normal load conditions to 2½% of the nominal voltage.

The Wiring Regulations also have no requirements as to the general appearance of an installation although in many cases an 'untidy' installation may indicate a possibility of bad workmanship. In this context it will be seen that the first regulation in Chapter 13, namely *Regulation 13-1*, requires that good workmanship and proper materials are to be used.

*Regulation 12-2* gives one of the ways in which the Wiring Regulations can have effect, namely by citation in their entirety in a contract, but they can also have effect:

(*a*) through a reference in statutory regulations or associated memoranda, as already explained;
(*b*) by citation in a British Standard;
(*c*) by citation as a condition of registration as an approved contractor;
(*d*) through reference in the requirements of licensing authorities.

The regulation also makes it clear that the Wiring Regulations cannot take account of every circumstance and there are cases where the Wiring Regulations will require to be augmented by specialised requirements depending on the nature of the installation concerned. An example of an installation of a special character is the hospital installation, in particular that part of such an installation which is in what the relevant British Standard calls the 'patient environment'.

The British Standard dealing with safety of medical electrical equipment is BS 5724, and its Part 1 covers principally general requirements but some installation requirements are included although it is understood that the latter are intended to be dealt with separately. As the work on BS 5724 progresses, it is also understood that the Department of Health and Social Security may refer to that Standard and discontinue their own 'Safety code for electromedical apparatus' (Technical Memoranda No. 8).

*Regulation 12-3* refers to the inter-relationship between the Wiring Regulations and statutory legislation. Although this takes various forms, one point which can be made immediately is that in the United Kingdom there is no legislation which *demands* that an electrical installation must comply with the IEE Wiring Regulations but in some cases compliance with them is deemed to satisfy the relevant legal requirements.

The Electricity Boards, in providing a supply to installations, are governed by the Electricity Supply Regulations 1937 and they are forbidden by those Regulations from 'permanently connecting a consumer's installation to the supply unless they are reasonably satisfied that the connection, if made, would not cause a leakage from the consumer's installation exceeding one ten-thousandth part of the maximum current to be supplied to the said installation'.

Another clause in the Electricity Supply Regulations states that the supply undertaking is not compelled to commence or continue to give a supply, unless they are reasonably satisfied that the consumer's installation meets certain safety requirements (which are prescribed in those Regulations). If the consumer's installation complies with the IEE Wiring Regulations it is deemed to satisfy those safety requirements of the Electricity Supply Regulations but as already inferred, compliance with the IEE Wiring Regulations is not the only way of satisfying the statutory requirements.

To be more precise, the IEE Wiring Regulations in this particular context mean the fundamental requirements for safety prescribed in Chapter 13. This relationship between these fundamental requirements and statutory regulations is stated in *Regulation 12-3* which also makes it clear that Parts 3 to 6 set out in greater detail the methods and practices which are regarded as meeting the fundamental requirements of Chapter 13 and hence also meet the safety requirements of the Electricity Supply Regulations and other statutory legislation.

The relationship between the IEE Wiring Regulations and the Electricity (Factories Act) Special Regulations 1908 and 1944 is somewhat different because in this case the former do not enjoy a 'deemed to satisfy' status but in the Memorandum on the Electricity Regulations (Booklet SHW 928, published by HMSO) it is stated that the IEE Wiring Regulations constitute a useful guide to safe construction and installation.

Those who are familiar with the Electricity (Factories Act) Special Regulations 1908 and 1944 will be aware of the fact that there are a number of Exemptions to those Regulations. For instance, if the system is such that the voltage does not exceed 125 V a.c. (or 250 V d.c.), a considerable number of the individual regulations are not applicable but the installation is still required to be, to quote, 'so constructed, installed, protected and maintained as to prevent danger so far as is reasonably practicable'. For this case, the Wiring Regulations are still a useful guide to construction and installation and therefore may still be applied even though the installation is not required to comply with the whole of the Electricity Special Regulations.

Of particular interest is Exemption 4 to the Electricity Special Regulations which gives total exemption to 'any process or apparatus used exclusively for electro-chemical or electro-thermal or testing or research purposes, provided that such process be so worked and such apparatus so constructed and protected and such special precautions taken as may be necessary to prevent danger'. Here again there is no reason why the Wiring Regulations cannot be used as the basis for designing such installations or parts of installations. For instance, the protective measures against electric shock detailed in Section 413 called 'protection by non-conducting location', 'protection by earth free local equipotential bonding' and 'protection by electrical separation' may be of interest in test areas.

*Regulation 471-23* of the Wiring Regulations is unique in that it is the only one which prescribes compliance with one particular regulation of the Electricity Special Regulations, namely Regulation 17, which deals with the dimensions of passageways and working platforms for open-type switchboards and other equipment. *Regulation 471-24* also makes specific mention of the Electricity Special Regulations in relation to areas accessible only to skilled persons, and it will be found that there are a number of other regulations in the Wiring Regulations which make reference to skilled persons and instructed persons.

With the advent of the Health and Safety at Work etc. Act 1974, it was decided to make new Regulations to replace the Electricity Special Regulations, and these will apply not only to factories but to all employment situations such as hospitals, teaching establishments, research premises, theatres, hotels, offices, shops and railway premises. Until these new Electricity Regulations are available it is, of course, not possible to comment on how they will relate with the IEE Wiring Regulations.

Buildings in Scotland, with certain specified exemptions, are required to comply with the Building Standards (Scotland) Regulations 1971 to 1979. Part N of those Regulations is concerned with electrical installations and this does not specifically require that an electrical installation must comply with the IEE Wiring Regulations but compliance with the latter is deemed to satisfy most of the Part N requirements.

Part N also includes additional requirements for installations in bathrooms, and there are separate requirements elsewhere in the regulations for lighting of exit routes and the provision of socket outlets in homes. It should be noted that the Scottish regulations at present refer only to the Fourteenth Edition of the Wiring Regulations, but it is open to local authorities, so it is understood, to accept compliance with the Fifteenth Edition for building regulation purposes in advance of amendment of the Scottish regulations to take account of the Fifteenth Edition.

Sometimes it is found that the relationship between the IEE Wiring Regulations and statutory regulations is an indirect one. An example of this concerns caravan sites which, whether they are permanent residential caravan sites or holiday caravan sites, are subject to the Caravan Sites and Control of Development Act 1960 (see also the Caravan Sites Act 1968). The enforcement body is the local authority of the area in which the site is situated, and the Secretary of State for the Department of the Environment may from time to time specify Model Standards concerning the layout of sites and the provision of facilities, services and equipment for these sites. Model Standards represent the standards normally to be expected, as a matter of good practice, and due account of them has to be taken by the local authorities when they formulate their own licensing conditions. The Model Standards revised in 1977 require that permanent residential caravan sites shall be provided with an adequate electricity supply and that the site installation shall be in accordance with the IEE Wiring Regulations.

There is a close liaison with the various government departments which are responsible for the statutory regulations and the final draft of a new edition of the

IEE Wiring Regulations is always submitted to the appropriate Secretaries of State for their opinion as to the acceptability of the new edition in relation to the statutory regulations. This is one example of the efforts which are made to see that there is no conflict between the IEE Wiring Regulations and relevant statutory regulations.

Mention must also be made here of the Electrical Equipment (Safety) Regulations 1975 (and the amendment to those Regulations published in 1976) because they cover any electrical equipment designed or suitable for household use and therefore cover not only electrical appliances, portable tools and luminaires but also wiring accessories (although some accessories are exempted from certain of the requirements). The scope of these Regulations is limited to equipment intended for use with a supply of not less than 50 V nor greater than 500 V a.c., or less than 75 V nor greater than 750 V d.c.

The 'Administrative Guidance on the Electrical Equipment (Safety) Regulations', issued by the Department of Trade primarily to aid the Local Authorities who are responsible for enforcing the Regulations, states that, although the Regulations do not refer to relevant national or international standards, 'it may reasonably be assumed that electrical equipment which has been manufactured to the safety requirements of acceptable standards, whether national or international, complies with the prescribed statutory requirements. Similarly the marks of recognised approvals bodies are reliable indication of compliance with the declared safety standards'. The Administrative Guidance then goes on to list the standards which, in the opinion of the Department of Trade, offer the degree of safety required by the Electrical Equipment (Safety) Regulations but then makes it quite clear that only the Courts can give authoritative rulings on points of law. Specific mention is made in the Administrative Guidance of equipment covered by certificates issued under the IEE Assessment of New Techniques Scheme.

Like the other statutory regulations mentioned earlier, the Electrical Equipment (Safety) Regulations are being revised, but until they are it is not possible to comment on what effect the revision may have on the relationship between these Regulations and the IEE Wiring Regulations.

The Electrical Equipment (Safety) Regulations are of particular interest because they were the first case of implementation by the United Kingdom government of a Directive of the European Economic Community concerning electrical safety. The Directive's full title is 'Council Directive of 19th February 1973 on the Harmonization of the Laws of Member States relating to Electrical Equipment designed for use within certain voltage limits' so it should come as no surprise that it is commonly called the 'Low Voltage Directive'.

The objects of this Directive are, firstly, that electrical equipment placed on the market shall not endanger the safety of persons, domestic animals or property when properly installed and maintained and used in applications for which it is made, and secondly, that free movement of all such equipment shall not be impeded.

The Low Voltage Directive covers all electrical equipment designed for use with a voltage rating of between 50 and 1000 V a.c., or between 75 and 1500 V d.c.,

other than
(i) equipment for use in explosive atmospheres
(ii) equipment for radiology and medical purposes
(iii) electrical parts for goods and passenger lifts
(iv) electricity meters
(v) plugs and socket outlets for domestic use
(vi) electric fence controllers
(vii) specialised electrical equipment for use on ships, aircraft or railways.
The Low Voltage Directive also excludes requirements concerning radio-electrical interference, these being the subject of another Directive, as are a number of the above items.

In addition to the close inter-relationship, as described earlier, between the IEC and CENELEC (both of which are voluntary standards bodies), there is another and extremely important factor in the participation of the United Kingdom in the work of CENELEC because one of the main aims of that body is the preparation of Harmonization Documents (and European Standards) to assist in the implementation of the Directives of the European Economic Community. An ECC Directive to the Member States will often need reference to technical requirements and although such requirements may be given in detail in the Directive, there is the possibility that the Directive can make reference to a CENELEC Harmonization Document (or European Standard) if this is considered adequate to meet their requirements.

Thus, when the Community accepts a CENELEC Harmonization Document dealing with the safety requirements for electrical equipment coming within the scope of the Low Voltage Directive, equipment complying with the national standard (of any Member State) corresponding to that HD must be regarded by the authorities in the other Member States as complying with the requirements of the Directive and cannot be refused entry to any country of the Community on the grounds of safety.

Before leaving the subject of the inter-relationship between the Wiring Regulations and statutory legislation the point must be made that designers, installers and owners of installations are responsible for establishing whether their particular installations are subject to statutory regulations, and they are also responsible for seeing that their knowledge of those regulations is up-to-date. Owners and users of installations have responsibility for seeing that those installations also continue to comply with the relevant statutory legislation and Chapter 34 of the Wiring Regulations concerning maintainability is an important one in relation to this aspect of continuing responsibility.

*Regulation 12-7* indicates that where the installation is in premises subject to licensing, the requirements of the licensing authority must be ascertained and, as with statutory legislation, the designer, the contractor and the user of the installation have the responsibility of determining these. For instance, the Greater London Council issue Technical Regulations for places of public entertainment which include requirements concerning the electrical installation in such places. The GLC Technical Regulations, in general, require compliance with the IEE Wiring

Regulations but also include some departures from, and additions to, the latter. As with statutory regulations, it is the responsibility of all concerned with an installation in premises subject to licensing to see that they have the latest information on any licensing requirements demanded by the authority concerned.

Because of the radical changes in approach and style used for the Fifteenth Edition the Wiring Regulations Committee decided that there was a need for a reasonably extended period of time where the Fourteenth and Fifteenth Editions should run in parallel, i.e. have equal validity. In arriving at their decision the Committee was particularly conscious of the need to give an adequate period for training and education authorities to introduce appropriate courses into their curricula and they decided that the date stated in *Regulation 12-9* did give that adequate period. During this moratorium period the designer can opt to use either edition but whichever one is used its requirements must be observed in their entirety and it is not permissible to mix the requirements of the two editions.

## 2.3 Chapter 13: Fundamental requirements for safety

Little needs to be included in this Commentary as regards Chapter 13, as the subsequent chapters of the Wiring Regulations are themselves in the nature of a description of methods of compliance with the fundamental requirements in Chapter 13, as indicated in *Regulation 12-3*. Because Chapter 13 is related to statutory legislation, it is written in broad terms which allow room for interpretation in particular cases.

As such interpretation is usually troublesome, attention tends to be focussed on the subsequent chapters which are designed for practical application in the great majority of installations so that Chapter 13 is normally invoked only where it is intended to adopt a practice not recognised in the later chapters (requiring the advice of a suitably qualified professional engineer, as demanded by *Regulations 12-2, 12-3* and *12-5*). However, as these later chapters are more comprehensive than previous editions as to the types of installation and methods of protection which are covered, there should be less need to invoke Chapter 13 in the manner just described. Chapter 13 is also used where legal action is contemplated in case of dispute and then it is up to the Courts to interpret the corresponding statutory regulations applicable to the particular case.

In gauging the significance of the phrase 'so far as is reasonably practicable' used in several of the regulations in this chapter of the Wiring Regulations, it should be borne in mind that the methods described in the later chapters must be regarded as reasonably practicable in most cases and should only be departed from where an unusual practice is proposed which makes certain requirements of the later chapters unnecessary or inappropriate. If the proposed practice results in a degree of safety less than that provided by compliance with the later chapters it is unlikely to be judged in the Courts or elsewhere that the designer has complied with Chapter 13 'so far as is reasonably practicable'.

Chapter 3

# Assessment of general characteristics

The main purpose of Part 3 of the Wiring Regulations is to indicate the characteristics of the source of energy and of the installation itself which are required to be known or assessed by the designer before he can proceed, but Section 314 consists of requirements concerning the subdivision of an installation, Chapter 33 concerns compatibility and Chapter 34 embodies requirements relating to the maintainability of an installation.

Part 3 must not be regarded as being mainly a check list for the designer because it cannot be emphasised too strongly that a full assessment in accordance with this Part is essential to the complete identification of the particular requirements of the Wiring Regulations which will be appropriate to the installation concerned.

## 3.1 Chapter 31: Purposes, supplies and structure

Whereas *Regulation 311-1* requires that the maximum demand of the proposed installation be assessed, *Regulation 311-2* does not mandatorily require that diversity be taken into account but the economic design of any installation will almost always mean that diversity cannot be ignored. However, it is essential that diversity factors are not over-optimistic, and where there is doubt as to the factors to be used an adequate margin for safety should be allowed. In any event, consideration should also be given to the possible future growth, and hence of maximum demand, of the installation.

As indicated in the Note to *Regulations 311-1* and *311-2*, Appendix 4 of the Wiring Regulations gives some information on the current demands for various points of utilisation and current-using equipment, together with some typical diversity factors which can be applied to, at least, the smaller installation.

In previous editions, diversity could not be applied to final circuits but that prohibition no longer exists in the Wiring Regulations except that for the standard circuit arrangements given in Table 5A of Appendix 5 diversity has already been taken into account and no further diversity is permitted.

*Regulation 312-3*, or more precisely Note 2 to that Regulation, indicates that, in determining the type of earthing arrangement(s) to be adopted, the designer may find it necessary to take account of the characteristics of the source of energy, particularly the facilities for earthing. This, of course, will always be the case when the installation is to be fed from the public supply network.

It is understood that it is the declared aim of the Electricity Supply Industry to make the necessary modifications to their low voltage distribution network in order that all consumers may be offered an earthing terminal connected by a continuous metallic return path to the earthed point of the source of energy. Thus in the great majority of cases, an installation together with its associated source of energy will constitute (as already), either a TN-S system, or where the supply employs Protective Multiple Earthing, a TN-C-S system. (See Fig. 3)

Some consumers who are supplied from a PME network may not be able to comply with the requirements of the PME Approval, and in such cases the designer of the installation has to provide another means of earth fault protection such as a residual current device requiring an installation earth electrode, the installation and source of energy then becoming a TT system. It is understood that the supply undertakings in these cases actually recommend the use of a residual current device.

It has already been pointed out in this Commentary that a complete installation may, in some cases, have to be considered as a collection of separate installations in respect of certain regulations. For instance, even where an earthing terminal is provided by the supply undertaking, and the combination of the 'general' installation and the source of energy is a TN-S system, if only part of that installation is to be protected by a residual current device and it is found necessary to use for that part an earth electrode, the part and the source are then treated as a TT system and the regulations applicable to an installation forming part of a TT system apply to that part but for the remainder of the installation those applying to TN systems must be met. In many instances in practice it will be found that such admixtures of types of system occur in relation to a particular installation.

*Regulation 313-1* is an important requirement even though items (i), (ii) and (v) may appear to be self-evident. In order to apply the requirements concerning protection against short circuit it is essential that the prospective short circuit current [item (iii)] at the origin of the installation is known, together with the type and rating, including the breaking capacity [item (iv)], of the overcurrent protective device at that origin.

In the larger installations where the supply is derived from a stepdown distribution transformer owned and controlled by the user, the prospective short circuit current is readily determined from a knowledge of the internal impedance of the transformer and the cable impedances. Where the installation is fed directly from the supply undertaking's low voltage distribution network, items (iii) and (iv) can be ascertained only from the supply undertaking itself. The supply undertaking may be able to indicate only that the prospective short circuit current will not exceed a certain maximum value rather than quote a particular value if for no other reason than that there could be a possibility of changes occurring in their network to cater, for instance, for changes in the load demand in the area concerned.

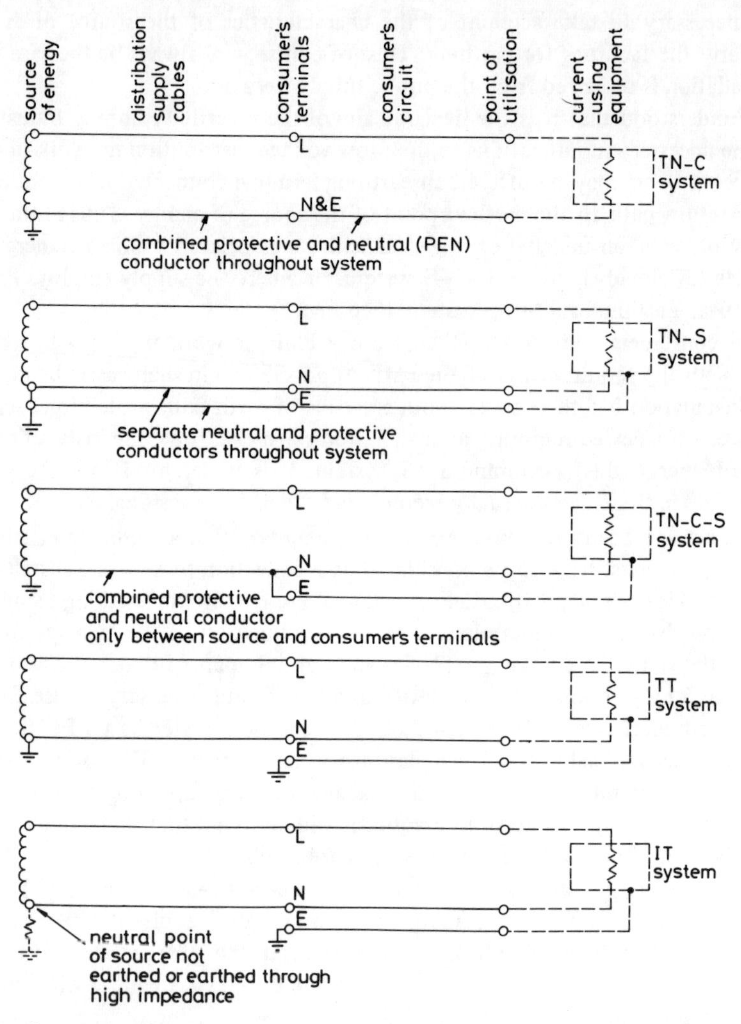


**Fig. 3** *System classification*

Between the earlier editions of the Wiring Regulations and the Fifteenth Edition there is a significant difference, in that the latter specifies in Section 413 the disconnection times for compliance with the requirements for protection against indirect contact when automatic disconnection of the supply is the protective measure used. This means a more careful estimation than hitherto, particularly where the devices concerned are overcurrent protective devices, of the earth fault loop impedances throughout the installation, and this cannot be undertaken without a knowledge of the contribution to those earth fault loop impedances of the source of energy and the supply cables external to the installation. Hence item (vi) of *Regulation 313-1*.

When the supply is already available, this impedance can be measured, or the designer or contractor may have some previous experience of similar installations in the same locality. When the supply is not in existence at the design stage, the information required can be obtained only from the supply undertaking, who again can give only a range of possible values or an expected maximum value. In the same way as already indicated in the case of the prospective short circuit current, changes in the supply system to cater for changes in load requirements may also modify the network's contribution to the earth fault loop impedances.

Appendix 3 of the Wiring Regulations gives explanatory notes on the types of system earthing and of the nomenclature used, so there is no point in repeating that information here although it is worthwhile reiterating the very last paragraph of those notes. That paragraph makes it clear that it is the responsibility of the consumer to satisfy himself that the characteristics of the earth fault current path, including any part of that path provided by a supply undertaking, are suitable to ensure operation of the type of earth fault protection used in the installation.

The Wiring Regulations of course cannot *demand* any appropriate action to be taken by the user of the installation to determine whether there has been any change in the earth fault loop impedances over a period of time and can only *recommend* periodic inspection and testing, the recommended frequency of such inspection and testing being given in Appendix 16 of the Wiring Regulations. Where the premises concerned are the subject of licensing, the licensing authority may dictate the frequency of testing and inspection.

*Regulation 314-1* indicates that the designer has to carry out an assessment of the intended installation to determine what subdivision of circuits is necessary to meet this regulation. The assessment will frequently and inevitably be of a subjective nature, mainly due to the fact that the definition of 'danger' includes injury to persons additional to that caused by shock and burns and the term 'inconvenience' is itself very subjective.

No hard and fast rules can be laid down which could be applied universally, each installation having to be treated on its own merits, but it would be reasonable to expect, for instance, that all the fixed lighting of an installation should not be fed by one final circuit, and that lighting circuits should be kept separate from power circuits.

*Regulation 314-3* is the first to mention the standard circuit arrangements detailed in Appendix 5, and it has to be emphasised that the adoption of a standard circuit arrangement allows the designer to assume that only some, not all, of the requirements of the Wiring Regulations have been met without further proof. He must, for instance, still assure himself that the chosen circuit is adequate for the expected load and meets the requirements of Chapter 41. If the overcurrent protective device for the circuit concerned is also intended to give protection against indirect contact, as will frequently be the case, he must see that the earth fault loop impedance does not exceed that giving disconnection in the time specified in Section 413 of the Wiring Regulations. The physical length of the circuit may well be limited by the magnitude of that part of the earth fault loop impedance on the supply side of the circuit concerned which, in turn, will include the contribution made by the impedances of the source of energy and the supply cables external to the installation. Neither does the use of a standard circuit arrangement permit the designer to ignore the limitation of voltage drop under normal load conditions prescribed in *Regulation 522-8* and this limitation may prove to be the factor which determines the circuit length and not the earth fault loop impedance requirement.

The use of one of the standard circuit arrangements is considered to give compliance with *Regulation 433-2* as regards overload protection, but it is still necessary to check that the rated breaking capacity of the overcurrent protective device is not less than the prospective short circuit current at the point of installation of the device. If this is the case, then *Regulation 434-5* allows the designer to assume that the requirements regarding short circuit protection are also met. If not, then *Regulation 434-6* would have to be applied, even though a standard circuit arrangement is involved.

*Regulation 314-4* combines two regulations of the previous edition of the Wiring Regulations and amongst other things does not allow the 'borrowing' of a neutral from another circuit.

## 3.2 Chapter 32: External influences

No Chapter 32 exists at present in the Wiring Regulations because, as pointed out, the IEC work is not yet sufficiently developed concerning the requirements for the application of the classification of external influences.

The classification itself is detailed in Appendix 6 of the Wiring Regulations and an examination of that Appendix gives an indication of the wide range of external influences it is intended to cover. The development of this classification is an important part of the aim of IEC Technical Committee 64 to arrive at reasonably concise rules with as near a universal application as possible, the intention being, as stated in the Note to Chapter 32, to employ the classification to distinguish between different types of installation and location and hence establish which requirements are relevant to a particular case or, for instance, which types of equipment would be admissible.

It should also be noted that there is a correlation between the IP Classification of BS 5490 'Degrees of protection provided by enclosures', and classifications 'AD' and 'AE' in Appendix 6 dealing with presence of water and presence of foreign solid bodies, respectively. Having identified the class within each external influence, e.g. AE2 and so on, relevant to the particular application concerned, the designer will then be able to determine the minimum IP classification for any equipment used in that application. Appendix B of this Commentary gives some details of the IP Classification.

## 3.3 Chapter 33: Compatibility

This chapter comprises only one regulation, namely *Regulation 331-1*, and this requires the designer to carry out an assessment of any characteristics of equipment likely to have an adverse effect upon other equipment or likely to impair the supply. The examples given are all of an electrical nature, some of which are also the subject of individual requirements in other chapters.

For instance, item (iii) draws attention to starting currents and *Regulation 552-1* requires that all equipment, including cables associated with a motor, shall be capable of carrying the starting and accelerating currents as well as the load current of that motor. Particular care is needed when the motor is intended for intermittent duty and for frequent stopping and starting. Transformers too, when switched on, can cause 'in-rush' currents several times larger than the normal load current.

Harmonic currents, referred to in item (iv), can among other things influence the cross-sectional area of neutral conductor which one can use, as indicated in *Regulation 522-9*, because the presence of significant harmonic currents, particularly third-harmonic currents (as they add up arithmetically in the neutral), will not permit the use of a reduced neutral, i.e. one having a cross-sectional area less than the associated phase conductors.

*Regulation 512-5*, which also concerns compatibility, makes it quite clear that this aspect is related to not only electrical characteristics, requiring that all equipment shall be selected and erected so that it will not cause harmful effects on other equipment. A number of regulations deal with compatibility between materials; for instance, *Regulation 523-10* deals with aluminium in contact with brass and Appendix 10 of the Wiring Regulations reiterates some notes which were in the Fourteenth Edition on protection against corrosion of exposed metalwork of wiring systems.

## 3.4 Chapter 34: Maintainability

The maintainability of an installation is an important factor in design. The maintenance which the installation in a private dwelling is likely to receive is, unfortunately, minimal. It is therefore necessary to allow for this fact in the quality of materials used and the protective measures to be selected because little reliance can be placed upon precautions to be taken by the user in the operation of the installation.

On the other hand, for an installation under the strict supervision of a competent authority, essential safety measures such as isolation can safely be left, at least partly, in the hands of the user. Furthermore, the use of equipment having a limited life is satisfactory if adequate arrangements exist for periodic replacement of the equipment.

There are, for example, two extreme attitudes to the selection of equipment to be used in situations subject to severe external influences, such as a corrosive atmosphere. The equipment can either be designed to withstand those influences for the foreseeable life of the installation, or, where that installation is under competent supervision, the equipment can be designed to have a deliberately limited life, as arrangements can be made for the equipment to be replaced whenever necessary. The latter alternative may be the more economic method in some cases.

Item (i) of *Regulation 341-1* which requires that periodic testing, maintenance and repairs can be readily and safely carried out does not mean that all equipment has to be accessible. However, the less accessible an item of equipment is, the greater must be its degree of reliability, unless its failure cannot cause danger.

As an example of this, *Regulation 526-1* requires joints in cables generally to be accessible for inspection, but excluded from that requirement are certain types of cable joint installed in enclosures having a suitable ignitability characteristic or embedded in building materials (such as structural concrete) for the life of the installation.

Chapter 4

# Protection against electric shock

Bearing in mind that the object of the Wiring Regulations is to provide safety, especially from fire, shock and burns, Part 4 can be regarded as probably the most important in the Regulations because it prescribes the protective measures which can be adopted and the circumstances in which those measures can be used.

Note 2 to *Regulation 400-1* (which applies to the whole of Part 4 and not only to Chapter 41) makes it clear that the order in which the protective measures are described does not imply any relative importance. Similarly, within each chapter, and this is particularly true of Chapters 41 and 43, the order in which the protective measures are described again does not mean one is more important than the other.

Mention has already been made in Chapter 1 of this Commentary of the fact that the order in which the various aspects of protection are treated, e.g. protection against electric shock before protection against overcurrent, does not necessarily coincide with the design sequence used by the designer. Nevertheless, after having carried out the assessments demanded by Part 3 of the Wiring Regulations, he will be able to identify the likely methods of protection against electric shock he intends to use and can establish from Section 471 if these are permissible, even if he does not check compliance with the relevant regulations in Chapter 41 until he has almost completed the design and checked, for instance, that it complies with Chapter 43 and with the voltage drop limitation under normal load conditions prescribed in *Regulation 522-8*.

Mention of Section 471 highlights the most important point to be remembered by the user of the Wiring Regulations, as stated in *Regulation 410-1*.

Chapter 41 merely catalogues the various types of protective measures against electric shock which are permissible but it cannot be used on its own. The designer must determine from Chapter 47, or more precisely Section 471, the circumstances in which the particular protective measure chosen can be applied. This point cannot be emphasised too strongly but for the purposes of this Commentary Chapter 41 and Section 471 are considered simultaneously and not separately. Chapter 53 prescribes requirements concerning the protective devices themselves and these will be considered later, but as Chapter 54 deals with requirements for the various protective conductors and is therefore an essential chapter when considering protection against indirect contact, comments on it are also included in this chapter of the Commentary.

Not all the measures listed in Chapter 41 are applicable to every installation or to every part of a particular installation; some of the measures are reserved for use only in particular locations and this fact is indicated in the appropriate regulations in Section 471, and will be considered in this present chapter of the Commentary.

However, before considering protection against electric shock, because it is in relation to this aspect that many of the new terms referred to earlier are to be found, a few brief comments on those terms seems advisable.

## 4.1 'Direct contact' and 'indirect contact'

The IEC Standards recognise a principle which has long been established in the IEE Wiring Regulations, namely that there should be two lines of defence against electric shock. The first line of defence is to give protection to the user in the normal working of the installation, for example, by providing basic insulation of equipment and the second to give protection in the event of a failure of that insulation, for example, by earthing the metallic enclosure of the equipment. In the IEC work, and adopted for the Fifteenth Edition, these two lines of defence can be identified with 'protection against direct contact' and 'protection against indirect contact', respectively.

## 4.2 Exposed conductive parts

The metallic enclosure of Class I current-using equipment is one example of an exposed conductive part because such equipment has only basic insulation and in the event of an earth fault caused, for instance, by a breakdown of that insulation, the metallic enclosure may become live. Metallic conduit, cable ducting and trunking are all further examples of exposed conductive parts, although such cable enclosures can be used as protective conductors and must then comply with the requirements for such conductors as prescribed in Chapter 54 of the Wiring Regulations.

An item of Class II current-using equipment may be encased in a metallic enclosure but because the equipment incorporates supplementary insulation (or reinforced insulation) in addition to basic insulation, the enclosure is not considered likely to become live in the event of a fault and is therefore not considered to be an exposed conductive part. However, as indicated in *Regulation 471-18*, if, for instance, a p.v.c.-insulated and p.v.c.-sheathed cable is installed in a metallic enclosure, this is not considered to be Class II construction – neither is a p.v.c.-insulated but unsheathed cable installed in an insulating enclosure such as p.v.c. conduit. Nevertheless, as also indicated in that regulation these combinations are considered, if correctly installed, to give satisfactory protection against both direct and indirect contact.

## 4.3 Extraneous conductive parts

Extraneous conductive parts are, by definition, conductive parts not forming part of the electrical installation but which are liable to transmit a potential and *Regulation 413-2* lists those which are commonly met in practice: main water pipes, main gas pipes, other service pipes and ducting, risers of central heating and air-conditioning systems and exposed metallic parts of the building structure.

## 4.4 Protective conductor

This is the generic term for conductors, not necessarily cable conductors, which are used in some of the measures for protection against indirect contact, the particular types being:

(*a*) circuit protective conductors
(*b*) combined neutral and protective conductors
(*c*) earthing conductors
(*d*) main equipotential bonding conductors
(*e*) supplementary bonding conductors
(*f*) earth-free equipotential bonding conductors.

Fig. 4 shows in somewhat more pictorial form the information given in Appendix 13 of the Wiring Regulations.

A circuit protective conductor connects the exposed conductive parts of equipment to the main earthing terminal of the installation and may be part of the same cable as the associated live conductors (*Regulation 543-5*), the metallic sheath and/or armour of that cable (*Regulation 543-8*), an enclosure of a factory-built assembly or of a busbar trunking system (*Regulation 543-7*), or rigid conduit, trunking or ducting (*Regulation 543-9*). The one type of circuit protective conductor which is not specifically mentioned, but is of course permissible, is a separate conductor such as that of a single core cable. As already mentioned, where enclosures of factory-built assemblies and busbar and cable enclosures are not used as circuit protective conductors, they are exposed conductive parts and sometimes they can have the two identities simultaneously – for one circuit being its protective conductor but an exposed conductive part relative to another circuit.

The earthing conductor is the conductor which connects the main earthing terminal to the means of earthing that may be an earth electrode or an earthing terminal provided by the supply undertaking who are then responsible for the connection to the sheath of the incoming supply cable, or in the case of PME, to the neutral of that cable.

Main equipotential bonding conductors connect the main gas and water pipes to the main earthing terminal and, as indicated by *Regulation 547-3*, the connection should be made as near as practicable to the point of entry of the non-electrical

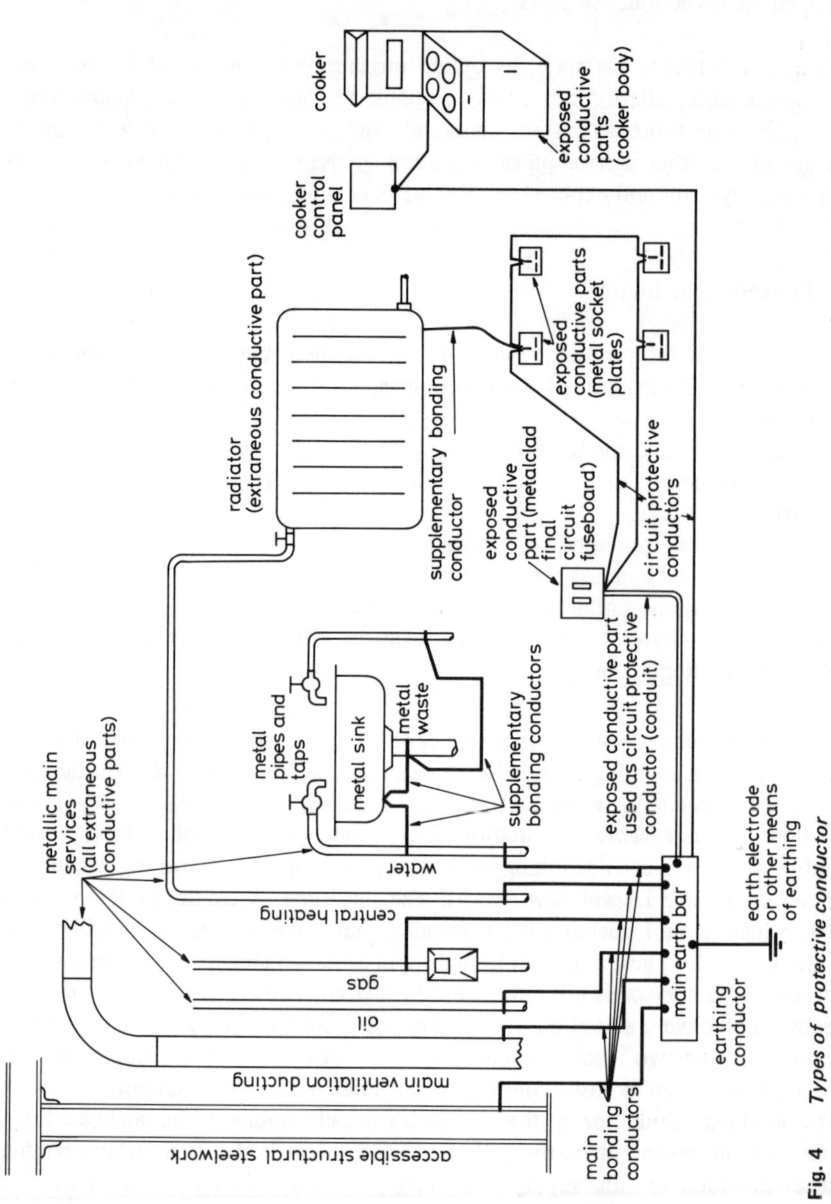


**Fig. 4** *Types of protective conductor*

Note: This Figure is not intended to illustrate requirements for bonding and earthing, but merely to identify the functions of particular protective conductors.

services into the premises concerned. As already indicated, other service pipes and ducting, risers of central heating and air-conditioning systems and exposed metallic parts of the building structure are also required to be connected to the main earthing terminal and the conductors used for this purpose are also main equipotential bonding conductors.

The basic requirement for supplementary bonding conductors is given in *Regulation 413-7*, and this regulation demands supplementary bonding of extraneous conductive parts only if they are simultaneously accessible with other conductive parts (either exposed or extraneous) and are not electrically connected to the main equipotential bonding by permanent and reliable metal-to-metal joints of negligible impedance. The regulation itself indicates that the purpose of this supplementary bonding is to maintain the equipotential zone created by the main equipotential bonding.

The Wiring Regulations now require supplementary equipotential bonding to be carried out in a room containing a fixed bath or shower, as prescribed in *Regulation 471-35*, even if there is no facility within the room for connecting a fixed item of current-using equipment. The reason for this requirement is that without such bonding of extraneous conductive parts, dangerous voltages could exist between those parts in the event of an earth fault elsewhere in the installation.

Supplementary bonding, as indicated in *Regulation 547-7* may be carried out by suitable extraneous conductive parts of a permanent and reliable nature on their own or in combination with bonding conductors, an example being shown in Fig. 5. For there to be a permanent and reliable nature means that there is no likelihood of plastics inserts or insulating plumbing materials being used in the metallic pipework concerned. Fig. 5 is considered in detail in Chapter 12 of this Commentary. Earth free equipotential bonding conductors are used in relation to the protective measure termed 'protection by earth free local equipotential bonding' and sometimes where the measure 'protection by electrical separation' is used.

Finally, in these brief notes concerning the new terms introduced into the Wiring Regulations, mention must be made of 'simultaneously accessible parts', because it is the existence of these parts which creates the shock hazard in an installation, of the type now termed 'indirect contact'. In the formulation of the rules for protection against electric shock due account has been taken, where necessary, of the possible introduction of simultaneously accessible parts into the installation *after* its completion and handing over by the installer. An example of this is the use of a non-conducting location (*Regulations 413-27* to *413-31*) which is limited by *Regulation 471-19* to special situations under effective supervision because without such supervision there would be the risk of the later introduction of 'earthy' conductive parts into the location thereby effectively destroying the protective measure and creating a dangerous situation.

Of course, in relation to the protective measure 'earthed equipotential bonding and automatic disconnection of the supply' simultaneously accessible parts (to persons or to animals) include exposed conductive parts of both fixed and portable equipment.

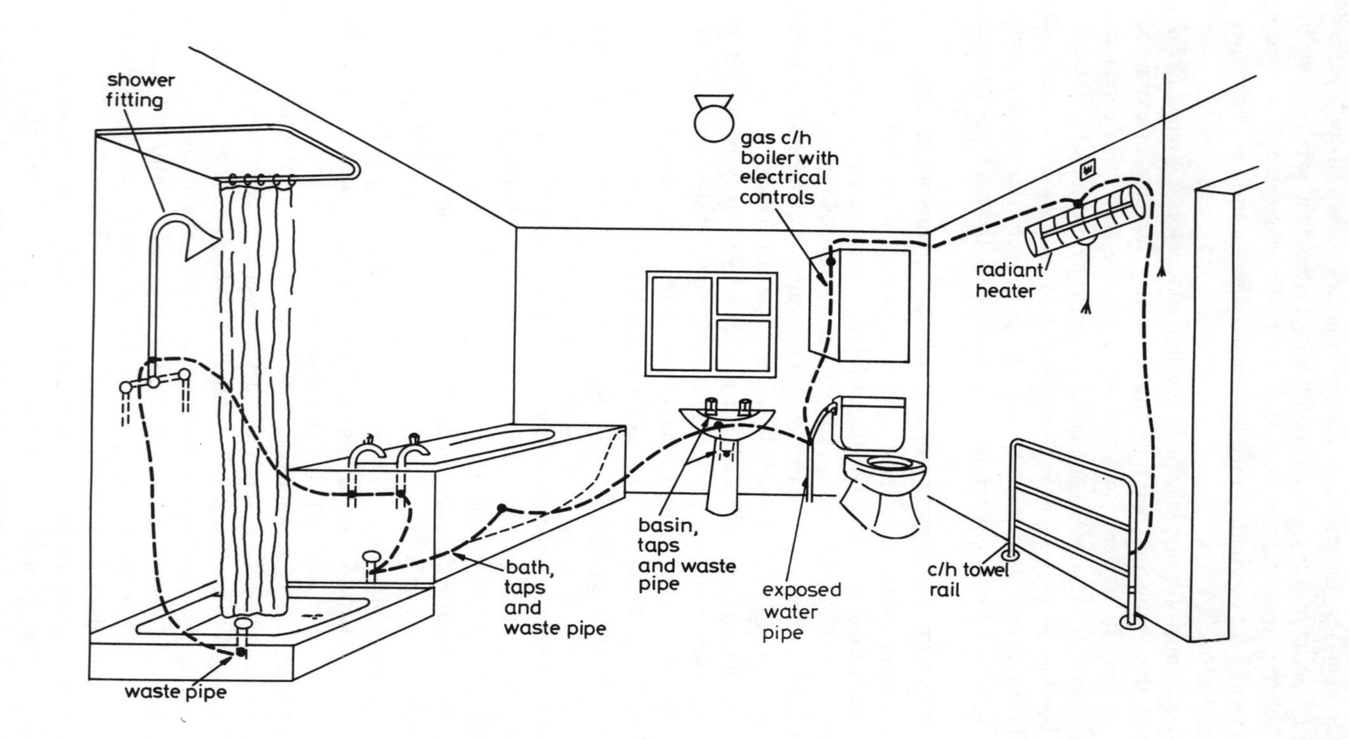


**Fig. 5** *Supplementary bonding*

The enclosure of the gas c/h boiler and the metallic parts of the radiant heater are exposed conductive parts. All other metallic parts are extraneous conductive parts.

Note: This diagram indicates supplementary bonding connections required between typical metallic parts in a bathroom. It is not intended to indicate practical routes of such connections.

## 4.5 Protection against direct contact

### 4.5.1 Protection by insulation of live parts

It will be seen that the first Section of Chapter 41 deals with measures giving protection against both direct and indirect contact, but as all three measures listed are of relatively limited interest they will be commented on later, and consideration given first to Section 412 dealing with protection against direct contact and then to Section 413 dealing with protection against indirect contact.

Fig. 6 shows in pictorial form a composite arrangement of examples of the protective measures prescribed in Section 412 of the Wiring Regulations for protection against direct contact.

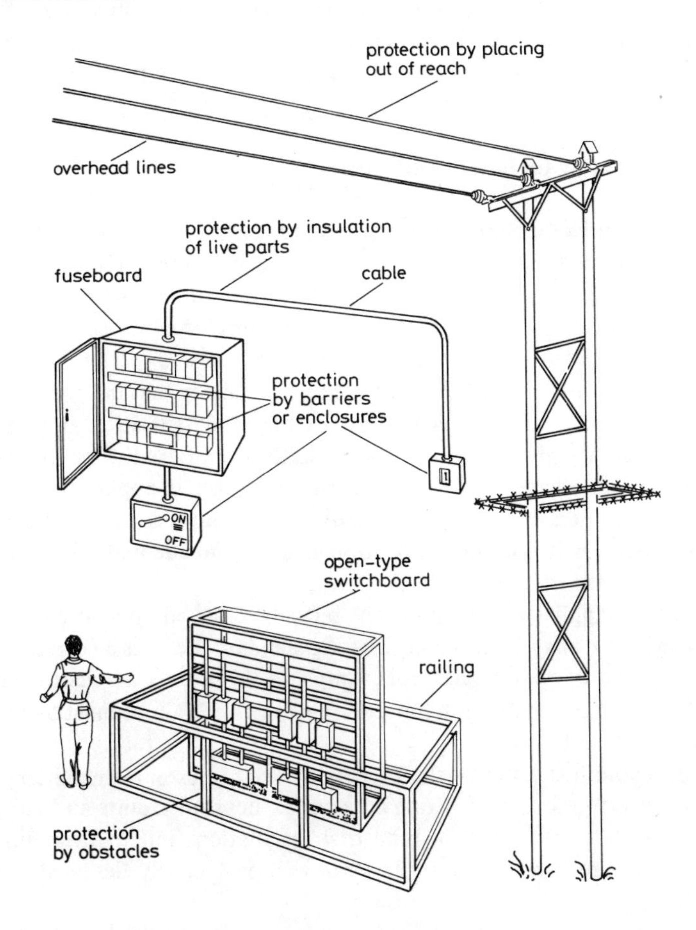


**Fig. 6** *Measures for protection against direct contact*

The first protective measure is 'Protection by insulation of live parts' as prescribed in *Regulation 412-2*, and this is usual in most installations, being provided by the normal insulation necessary for the working of the equipment concerned. Where this insulation does not provide protection against direct contact (as, for example, in the case of a bare conductor supported on insulators) the protective measure can be applied on site. However, when this is undertaken, the quality of the insulation applied by the contractor needs to be confirmed by tests similar to those which ensure the quality of insulation of similar factory-built equipment. As such tests usually necessitate the use of high-voltage test equipment and special methods such as the application of a foil electrode to the parts under test, this protective measure will be found to be impracticable by many contractors for use on site and the use of factory-built equipment will be preferred. Note 2 to *Regulation 412-2* reiterates what is found in many equipment safety standards, namely that paints, varnishes, lacquers and similar products do not on their own provide adequate insulation for protection against direct contact. Parts using such finishes therefore must not be accessible through apertures in the enclosure of the equipment concerned.

### 4.5.2 Protection by barriers or enclosures

The next protective measure for which requirements are given (*Regulations 412-3* to *412-6*) is termed 'protection by barriers or enclosures', although there is an essential difference between a barrier and an enclosure in that whereas the former is intended to prevent contact with live parts from any usual direction of access, the latter is employed to prevent such contact from any direction. A barrier, for instance, need not have a top surface if access from above that barrier is unlikely.

*Regulation 412-3* is one which mentions the IP classification and Appendix B of this Commentary gives details of that classification. Here the specified degree of protection IP 2X means that if a standard test finger is applied to apertures of enclosures or barriers it will not touch live parts, and if a 12 mm rigid sphere is applied to openings it does not pass through them and adequate clearance is maintained to live parts inside the enclosure.

The measure applies to all equipment but where the equipment complies with a British Standard it may be assumed that the standard includes a corresponding test for non-accessibility of live parts. An example of this is the consumer's control unit to BS 5486 Part 13, requiring that live terminals of such a unit must be inaccessible to the standard test finger.

For household installations, the barriers or enclosures one encounters are those which are an integral part of wiring accessories, consumer units and current-using equipment. This is also true of industrial installations but in these installations such features, together with obstacles, can well be specially designed and erected in order to obtain the protective measures.

It is worthwhile pointing out that in some British Standards, such as BS 3456 'Safety of household and similar electrical appliances', there is a requirement that there shall be adequate protection against the risk of contact with basic insulation but this insulation can be accessible after removing a cover that requires the use of tools for removal and has to be removed for user maintenance. If the cover can be removed without the use of tools a warning has to be given on that cover indicating that a plug has to be removed or the appliance disconnected from the mains before removing the cover.

A common example illustrating the exception stated in *Regulation 412-3* is given in *Regulation 471-8* and is the independent lampholder for which it is clearly not practicable to provide complete protection against direct contact when the lamp is removed. Apart from this one example, the exception will be found to be applicable only to special locations.

In *Regulation 412-4*, mention is made of another degree of protection in the IP Classification, namely IP 4X. This is a higher degree of protection than IP 2X and compliance is checked using a straight rigid steel wire 1 mm in diameter. The purpose here is to guard against contact which would occur if a metallic wire or strip greater than 1 mm thickness drops through or is pushed into an opening in a horizontal top surface of an enclosure.

As regards *Regulation 412-6*, the Electrical Equipment (Safety) Regulations do not require the use of key or tool for opening enclosures of ceiling roses, cord-operated ceiling switches or fixed wiring junction boxes even if, after removal of covers, live parts are accessible. This exemption was considered justified because such accessories are usually installed out of reach. Even for these accessories it is believed that compliance with *Regulation 412-6* can be met by a suitable design but for ceiling roses it was accepted that provided they complied with BS 67 they should be totally exempted from that regulation. It is believed that a similar exemption can be given to cord-operated ceiling switches.

The two protective measures already commented on can be used in, and will be found in, any installation, but as indicated in *Regulations 471-6* and *471-7* respectively, they must be associated with one or more of the protective measures for *indirect* contact. The two remaining protective measures in Section 412 ('protection by obstacles' and 'protection by placing out of reach') may be associated with one or more of the protective measures for indirect contact but their use, as indicated in *Regulations 471-9* and *471-10*, respectively, is limited to areas accessible only to skilled persons or to instructed persons where the latter are under direct supervision.

### 4.5.3 Protection by obstacles

*Regulations 412-7* and *412-8* prescribe the requirements for protection by obstacles and it must be emphasised that obstacles are not intended to prevent *intentional* contact with live parts. The measure is a precaution against accidental contact and should not be confused with 'live-working' which is admissible in locations accessible only to skilled persons.

It will be noted that a distinction has been made between locations reserved for skilled persons only and locations for skilled persons and instructed persons where the latter persons are under direct supervision. This has to be made in order to meet the requirements of the Electricity (Factories Act) Special Regulations 1908 and 1944. It will be seen from *Regulation 471-24* that if the location is for skilled persons only and access can only be gained by a safety ward lock key or tools, the Special Regulations in certain cases do not require any measures to be taken against either direct contact or indirect contact.

### 4.5.4 Protection by placing out of reach

The last of the protective measures given in Section 412 for protection against direct contact is termed 'protection by placing out of reach', the requirements for which are prescribed in *Regulations 412-9* to *412-13*. This protective measure is another which is limited to locations accessible only to skilled persons or instructed persons, as stated in *Regulation 471-10*, unless the conductors concerned are overhead lines for distribution between buildings in which case they must then comply with the Overhead Line Regulations.

The purpose of *Regulations 412-9* to *412-13* is self-evident and does not require comment here except to point out that 'arm's reach' must not be confused with simultaneous accessibility of conductive parts. Arm's reach as a concept is confined to consideration of protection against direct contact whereas simultaneous accessibility of conductive parts is the basic factor when dealing with protection against indirect contact.

With regard to *Regulation 412-13*, it is obviously not possible to specify the actual distances one should aim for in locations where bulky or long conducting objects are normally handled, as these distances will largely depend on the nature of the work carried out in these locations. The installation designer may not be aware of the intended use of the premises and in such circumstances it may be necessary to adopt an alternative protective measure against direct contact.

## 4.6 Protection against indirect contact

Under the Electricity Supply Regulations 1937 a supply undertaking, i.e. an Area Electricity Board, is required to earth the low voltage supply to installations at one point, usually the neutral point of the secondary winding of the distribution transformer at their substation. When Protective Multiple Earthing is used, the PME Approval modifies that statutory requirement and permits the supply undertaking to earth the supply at more than one point.

Thus, in the United Kingdom, as already indicated in Chapter 3 of this Commentary, an installation together with the source of energy will comprise either a TN-S system or, with PME, a TN-C-S system. When it is not possible for the supply undertaking to provide the necessary metallic path from the main earthing terminal of an installation to the earthed point of the source, the owner of the installation has to provide an earth electrode and the combination of source and installation is then a TT system.

Consideration here is therefore limited to these types of system and Appendix E deals with the remaining type (the IT system) in which the source of energy is not earthed or is earthed through a high impedance, because this latter type of system is of comparatively limited interest and, in any event, cannot be used where the installation is fed directly from the public low voltage supply network.

Section 413, as indicated in *Regulation 413-1*, details five protective measures which can be used to give protection against indirect contact, but four of these are also of limited interest, as will be seen later. The first measure, item (i) in *Regulation 413-1* ('protection by earthed equipotential bonding and automatic disconnection of supply') is the one most commonly used and *in principle* the requirements prescribed in the Fifteenth Edition are no different from those of preceding editions but there are important and significant additions, as will be seen.

Because of the extensive use of this protective measure it is considered here in some detail.

### 4.6.1 Protection by earthed equipotential bonding and automatic disconnection of supply

As demanded by *Regulation 413-2*, extraneous conductive parts are required to be connected to the main earthing terminal of the installation thereby creating an equipotential zone. The small installation such as the typical household installation will usually comprise only one such zone but in the larger installation there may well be a number of these equipotential zones. For instance, when the installation supplies a number of buildings, the main equipotential bonding requirements have to be carried out for each building at the point of intake, thereby creating a separate equipotential zone in each building.

Exposed conductive parts are also required to be connected by means of circuit protective conductors to the main earthing terminal, (or for the installation having a number of zones, to the earthing terminal at the point of intake of the zone in which the equipment concerned is situated) the pertinent regulations being *Regulation 413-8* when the installation is part of a TN system, or *Regulations 413-10* and *413-11* when it is part of a TT system.

The purpose of this creation of an equipotential zone, as indicated by Note 1 to *Regulation 413-2*, is to minimise the voltages to which the person protected may be subjected in the event of an earth fault in the zone. Appendix C of this Commentary analyses the magnitudes of these voltages and even when the person protected is within an equipotential zone it will be seen that these voltages can be dangerous if allowed to persist.

Thus the key requirement of this protective measure is the rapid disconnection of the supply in the event of an earth fault caused, for example, by the breakdown of basic insulation in an item of current-using equipment so that the phase conductor comes into contact with the exposed conductive parts of that equipment, the disconnection of the supply being effected by either the use of overcurrent protective devices, residual current devices or fault-voltage operated earth leakage breakers.

The existence of these voltages between exposed conductive parts in a zone in the event of an earth fault may suggest that the name 'equipotential zone' is a misnomer, but it must be remembered that the zone is truly equipotential in the event of an earth fault occurring in the supply cable to the zone, e.g. between the cable phase and protective conductors. In this case the main earthing terminal of the zone may acquire some potential above Earth but all the exposed and extraneous conductive parts connected to that terminal will also acquire the same potential and provided that the person protected remains within the zone there is no risk of electric shock.

When the earth fault occurs in the installation the Fifteenth Edition recognises that the degree of risk is greater with portable equipment held in the hand compared to that with fixed equipment and for this reason *Regulations 413-4* requires disconnection within 0·4 s for socket outlet circuits and 5 s for circuits feeding fixed equipment.

If for socket outlet circuits it is found that the limiting values of earth fault loop impedance given in Table 41A1 cannot be met, the alternative method (which is detailed in Appendix 7 of the Wiring Regulations) for compliance with *Regulation 413-3* permits one to extend the maximum permissible disconnection time to 5 s, and hence to use the higher limiting values of earth fault loop impedance given in Table 41A2, at the expense of having to limit the impedance of the protective conductor of the circuit concerned to the values given in Tables 7A and 7B. *Regulation 471-14* also points out that when it is not possible to meet the limiting values of earth fault loop impedance to enable overcurrent protective devices to be used the use of a residual current device is then preferred.

The Fifteenth Edition also recognises, in *Regulation 471-12*, that the degree of risk is greater if the person protected uses equipment outside the zone in which the circuit feeding that equipment originated, e.g. when that person uses equipment in the garden fed from a socket outlet in the house, because not only will he or she be in direct contact with the general mass of earth but may have a reduced body resistance, this latter aspect being referred to in *Regulation 471-11* and the note to that regulation.

Thus for the first time the Wiring Regulations introduces, in *Regulation 471-12*, a requirement that for a socket outlet rated at 32A or less, specifically intended to supply equipment where the conditions are as described in the previous paragraph, a residual current device having a rated residual operating current not exceeding 30mA must be used as the protective device. This requirement also applies if the equipment is fed from other than a socket outlet but still via a flexible cable or cord of 32A or less current-carrying capacity and the disconnection time must still comply with *Regulation 413-4* (i), i.e. must not exceed 0·4 s. Where

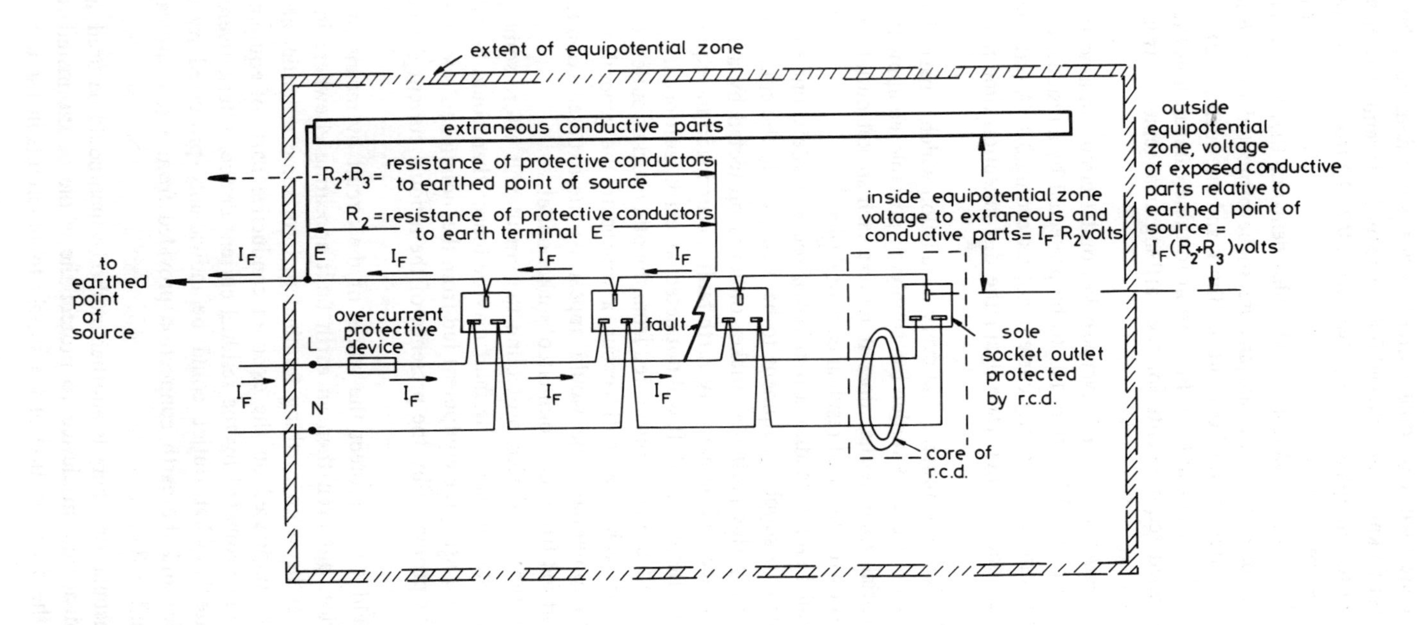


**Fig. 7** *Non-detection of earth fault by r.c.d. protecting only one socket outlet in multisocket outlet circuit.*
Socket outlets shown from rear

other equipment is concerned a residual current device is not demanded but the disconnection time must likewise not exceed 0·4 s. The reader is reminded that any socket outlet installed for compliance with *Regulation 471-12* has to be provided with a notice as required by *Regulation 514-8*.

The Fifteenth Edition, in *Regulation 471-44*, also demands that socket outlets intended to supply mobile (touring) caravans are to be protected by a residual current device having a rated residual current of 30mA, and one such device can be used to protect up to six socket outlets. In relation to these caravan sites, *Regulation 471-43* gives detailed requirements for the earthing arrangements which are acceptable for such sites.

As indicated by *Regulation 471-13*, every socket outlet circuit in a household installation which is part of a TT system has to be protected by a residual current device having a rated residual operating current not exceeding 30mA (one device can protect more than one circuit). Here again the disconnection time must not exceed 0·4 s.

*Regulation 471-12* does not require that every installation which is part of a TN system must incorporate a socket outlet for supplying portable equipment used outside the equipotential zone neither does it require each socket outlet for this application to have its own individual residual current device.

Fig. 7 shows a radial circuit feeding a number of socket outlets, one of which is specifically intended to supply equipment to be used outside the equipotential zone. If, as shown, only this particular socket outlet is protected by a residual current device, that device will detect only earth faults occurring in the equipment being supplied or in its flexible cord. It will not detect an earth fault occurring anywhere in the circuit itself or in equipment fed from the other socket outlets but, as shown in the Figure, if such a fault did occur, a voltage to the general mass of Earth outside the equipotential zone would appear on the exposed conductive parts of the equipment fed from the 'nominated' socket outlet.

The overcurrent protective device protecting the circuit will afford some protection but the time taken to clear the fault may be longer than could be safely tolerated and the person using the equipment fed from the 'nominated' socket outlet is not additionally protected by the presence of the residual current device as he may believe.

One solution would be to protect the whole of the circuit by means of the residual current device, but even then an earth fault appearing elsewhere in the installation will cause the main earthing terminal to have some potential above Earth and this will be impressed on the exposed conductive parts of equipment connected to the circuit protected by the residual current device. Alternatively of course, the 'nominated' socket outlet could be individually protected by the residual current device and the earth connection provided from a separate earth electrode, as shown in Fig. 8.

For the smaller installation, there is another method which could be used, and that is for one residual current device to protect the whole of the installation although this means the disconnection of the supply to all circuits in the event of

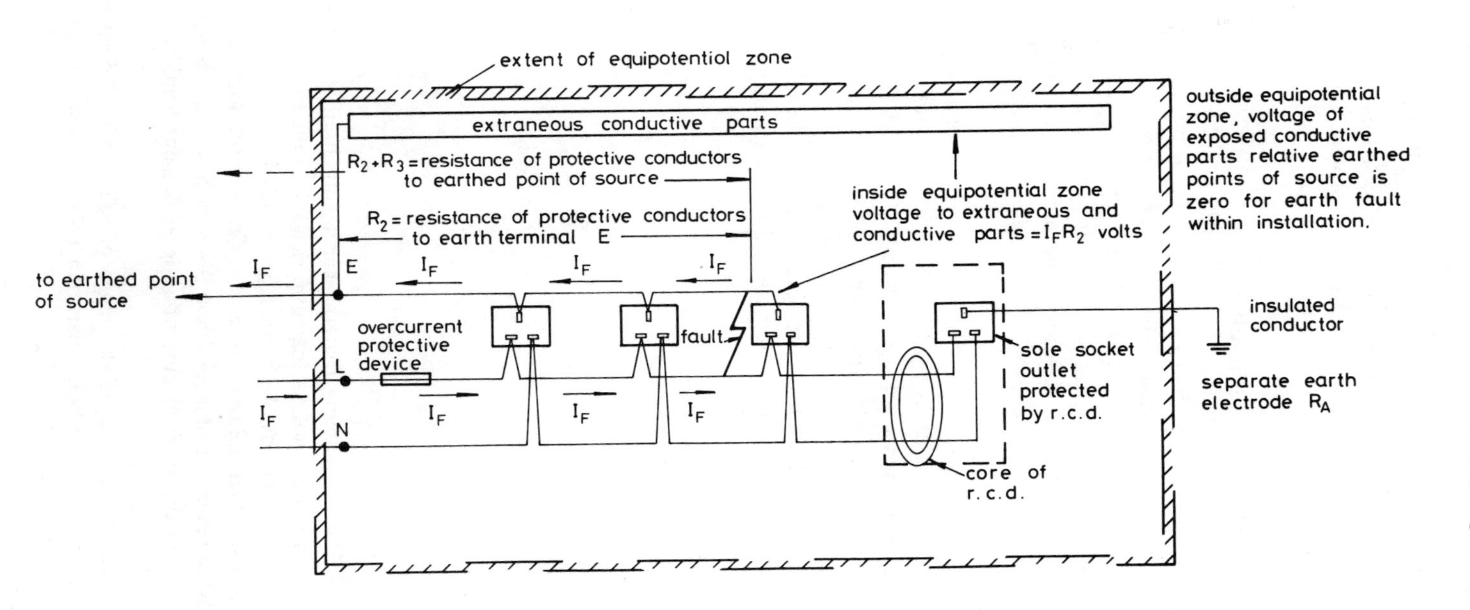


**Fig. 8** *Use of separate earth electrode for r.c.d. protecting only one socket outlet in multisocket outlet circuit*
Socket outlets shown from rear

an earth fault. Because of this it is suggested that a more reasonable solution would be to use one residual current device to protect all the socket outlet circuits and another such device or overcurrent protective devices to protect the other circuits of the installation, bearing in mind the requirement in *Regulation 314-1* concerning avoidance of inconvenience.

*Regulation 413-6* requires that when a residual current device is used, the product of its rated residual operating current in amperes and the earth fault loop impedance in ohms shall not exceed 50 (V). The purpose of this regulation is to ensure that in the event of an earth fault, sufficient residual operating current will be developed in order to obtain disconnection within the permitted time. It also ensures that when there is no earth fault, the presence of earth *leakage* currents from equipment supplied by the circuit being protected will not cause excessive voltages to occur on exposed conductive parts or between those parts and extraneous conductive parts.

Compliance with *Regulation 413-6* can be regarded as automatic in the case of an installation which is part of a TN system, where the rated residual operating current is 30 mA, and the nominal voltage to Earth, $U_0$, is 240 V. This can be readily shown in the following manner, where the rated residual operating current is denoted by $I_{\Delta n}$ A and the earth fault loop impedance by $Z_s$ Ω.

Then, in the event of an earth fault occurring in the circuit, the minimum earth fault current (for a fault occurring at the remote end of the circuit), $I_f$ amperes, is given by

$$I_f = \frac{U_0}{Z_s} = b\, I_{\Delta n} \text{ amperes}$$

but

$$I_{\Delta n} Z_s \leqslant 50$$

so that from these two equations

$$b \geqslant \frac{U_0}{50}$$

Thus, when $U_0 = 240$ V and assuming that the earth fault itself is of negligible impedance and that no part of the normal load impedance is in the earth fault loop, the earth fault current *must be* at least 4·8 times $I_{\Delta n}$ (i.e. $b \geqslant 4{\cdot}8$).

It must then be checked that the residual current device at this value of residual operating current has an operating time less than 0·4 s, or 5 seconds, whichever is appropriate, depending on the type of current-using equipment being fed by the circuit.

It is, however, instructive to indicate the range of values of '*b*' which will be obtained in practice when the installation is part of a TN system and when no separate earth electrode is used.

Assume that the impedance of that part of the earth fault loop impedance external to the circuit being protected is equal to or less than 0·5 Ω (which is not an unrealistic value), the circuit including its protective conductor is wired in p.v.c.-insulated cable to BS 6004, the circuit meets the limitation of voltage drop under normal load conditions as prescribed in *Regulation 522-8* and the design current of the circuit is in the range 6A to 32A. Then, when $U_0$ is 240 V and $I_{\Delta n}$ is 30 mA it can be shown that '*b*' will be in the range of values 340 to 1160.

Obviously, even if part of the normal load impedance because of the location of the earth fault becomes part of the earth fault loop, the fault current will still be considerably greater than 4·8 $I_{\Delta n}$.

The international proposals for residual current circuit breakers (see Chapter 10 of this Commentary) for household and similar use, and which are also under consideration for the revision of BS 4293, include the following table of maximum operating times.

**Table 1: Proposed maximum operating times for residual current circuit breakers**

| Rated residual operating current $I_{\Delta n}$ mA | Maximum total operating time in seconds | | | |
|---|---|---|---|---|
| | at $I_{\Delta n}$ | at $2I_{\Delta n}$ | at $5I_{\Delta n}$ | at 0·25A |
| 6 | 5 | 1 | - | 0·04 |
| 10 | 5 | 0·5 | - | 0·04 |
| 30 | 0·5 | 0·2 | - | 0·04 |
| greater than 30 | 2 | 0·2 | 0·04 | - |

If the system is TT so that a local earth electrode has to be used, as indicated in Fig. 9, '*b*' is again equal to or greater than $U_0/50$ and if $U_0$ is 240 V the earth fault current must be at least 4·8 $I_{\Delta n}\ R_A/Z_s$ amperes.

In this case the actual value of '*b*' obtained in practice will be very much less than with the TN system and will depend on what portion of the total earth loop impedance $Z_s$ is contributed by the resistance of the local earth electrode.

Before leaving the subject of residual current devices, at least for this present chapter, mention must be made of the Note to *Regulation 471-14*. This recognises the fact that a residual current device having the characteristics referred to in the Note, as well as giving protection against indirect contact, will also offer an increased degree of protection against fatal electric shock in the event of misuse of equipment or the failure of other protective measures.

If one examines the details of cases of electrocution, in the home for instance, there can be no doubt that lives could have been saved in the circumstances just mentioned if the circuit had been protected by a 30mA residual current device or one of higher sensitivity (i.e. having a lower rated residual operating current) but it must be recognised that the main aim in using any residual current device is to give protection against *indirect* contact.

The use of a residual current device as regards protection against direct contact must always be supplementary to the measures specified in items (i) to (iv) of *Regulation 412-1*, and there are no exceptions to this. Furthermore, the presence of a residual current device in an installation must never be regarded as a reason for neglecting to maintain that installation and the equipment it serves in a good condition.

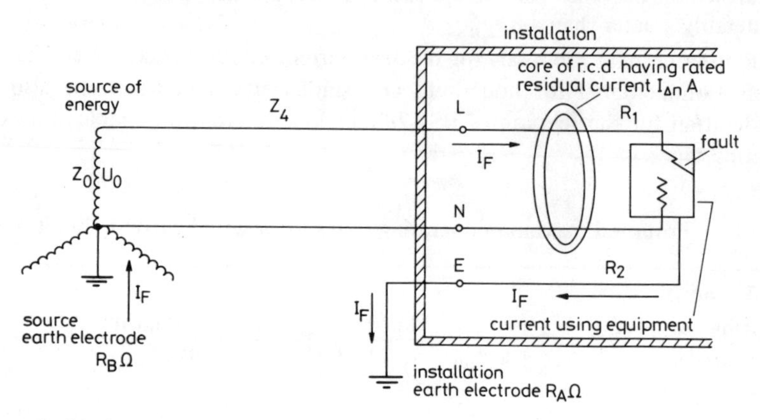


**Fig. 9** ***Residual current device in installation in TT system***

$$I_F = \frac{U_0}{Z_0 + Z_4 + R_1 + R_2 + R_A + R_B}\,\text{A} = \frac{U_0}{Z_s}\,\text{A} = bI_{\Delta n}\ \text{A}$$

but as $\quad R_A\, I_{\Delta n} \leqslant 50, \quad b \geqslant \dfrac{U_0}{50}\,\dfrac{R_A}{Z_s}$

Notwithstanding the increasing interest shown in residual current devices and the support given to their use by the Fifteenth Edition it is still true to comment that, in general, the main method of protection against indirect contact is to use the devices affording protection against overcurrents in the circuits concerned, and this has been recognised by the inclusion of *Regulation 413-5* and its associated tables giving limiting values of earth fault loop impedance both for 0·4 s and 5 s disconnection, which, as stated in *Regulation 471-12*, apply only within the equi–potential zone created by the main equipotential bonding. It will be noted that in these tables the limiting values of earth fault loop impedance where the device is a miniature circuit breaker are exactly the same for 0·4 s and 5 s disconnection. With these devices it was found necessary to base the values on the instantaneous tripping current which is defined in BS 3871 as the minimum current to cause tripping within 0·1 s. Further information on the practical implications of these tables is given in Appendix C of this Commentary, and it will be seen that in very many cases the limiting values of earth fault loop impedance are readily met when the installation is part of a TN system.

If the installation is part of a TT system it will usually be the case that the earth fault loop impedances obtained will be insufficiently low to give disconnection in the specified times using overcurrent protective devices to give protection against indirect contact and therefore a residual current device or devices may have to be used for circuits feeding fixed equipment as well as for socket outlet circuits. For the same reason it is most unlikely that the alternative approach detailed in Appendix 7 of the Wiring Regulations can be used in such installations.

In *Regulation 471-11*, mention is made of the fact that the limiting values of earth fault loop impedance specified in *Regulation 413-5* apply only to normal dry conditions and other, lower, values should be used where reduced or very low body resistance is to be expected. However, the Wiring Regulations do not specify these lower impedance values or give other information except in relation to circuits supplying equipment used outside the protected zone as already described in the comments on *Regulation 471-12* and in relation to rooms containing a fixed bath or shower where *Regulation 471-36* demands that fixed equipment (socket outlets other than shaver supply units are not permitted in such rooms) is disconnected in 0·4 s. Furthermore, in these rooms, in order to further minimise the voltages which would appear between simultaneously accessible parts, all such parts must be bonded together, i.e. supplementary bonding must be used.

It has been felt that more detailed regulations concerning the more onerous environments where reduced or very low body resistance is to be expected should await the outcome of the international discussions on this very important aspect.

As a matter of some interest, *Regulation 471-37*, which requires that the shaver supply units used in rooms containing a fixed bath or shower shall comply with BS 3052, represents a combination of two protective measures against indirect contact, namely, 'protection by electrical separation' and 'protection by the use of Class II equipment' because the standard requires that the secondary circuit supplying the socket outlets is isolated both from the incoming supply and Earth and also specifies precautions to prevent the use of the unit to supply equipment other than that intended.

Although the use of Class II fixed equipment is generally permissible in a room containing a fixed bath or shower, it should be checked that the equipment complies with an appropriate British Standard in such a way that the properties of its supplementary insulation are not likely to be impaired by subjection to the degree of moisture to be expected in the situation in which it is erected.

Another method of protection against indirect contact, but still within the measure known as automatic disconnection of the supply, which can be used in bathrooms or elsewhere where the person protected is expected to have reduced body resistance is to reduce the nominal voltage to earth. As will be seen later in this Chapter, the Wiring Regulations include requirements for three well defined reduced-voltage systems, but before commenting on these there are three remaining protective measures against indirect contact described in Section 413 which require consideration.

### 4.6.2 Protection by use of Class II equipment or by equivalent insulation

It is not for this Commentary to concern itself with the arguments which still persist among some electrical engineers on the relative merits of earthing and the use of double insulation as the means of protection against indirect contact. The Wiring Regulations admit both methods, the latter being termed 'protection by use of Class II equipment or by equivalent insulation' and the requirements for which are prescribed in *Regulations 413-18* to *413-26*. In the comments which follow, reference to Class II equipment relates to appliances and luminaires rather than to wiring accessories.

In Class II equipment, the first line of defence for protection against indirect contact is its basic insulation, and the second line of defence takes the form of reinforced insulation or, preferably, a separate layer of supplementary insulation. There is no provision for the connection of exposed metalwork (if any) to a protective conductor and no reliance is placed upon precautions to be taken in the fixed wiring of the installation.

Class II equipment may have a durable and substantially continuous enclosure of insulating material which envelops all metal parts except small parts such as nameplates, screws and rivets (but these small parts have to be isolated from live parts by insulation at least equivalent to reinforced insulation). Such an equipment is now called 'insulation encased Class II' but was earlier known as 'all insulated'. Another type of Class II equipment having a substantially continuous metallic enclosure has double insulation throughout (except that where such insulation is manifestly impracticable, reinforced insulation may be used). This latter type is termed 'metal encased Class II equipment' or 'double insulated equipment'. The metallic enclosure, as indicated earlier in this Chapter, is not considered likely to become live in the event of a fault and is not therefore an exposed conductive part.

If equipment has double insulation and/or reinforced insulation throughout but also an earthing terminal it is considered to be Class I and not Class II but the equipment standards allow a Class II equipment to be provided with means for maintaining the continuity of protective circuits not associated with the equipment itself, in the same way as stated in *Regulation 413-25*.

Metal parts of Class II equipment, in fact, should never be earthed because this would entail a careful check of the construction of the equipment to verify that it complied with all the requirements for Class I as the items of metalwork concerned may not be electrically continuous with other items of exposed metalwork of the equipment. Any attempt to modify the construction of a Class II equipment risks invalidating its compliance with the British Standard.

Class II construction is being increasingly used for many types of portable equipment and for some equipment, such as mains operated lawnmowers, Class II is the *only* method of construction permitted by British Standards. However, the designer of the fixed installation is not generally concerned with the selection of such portable equipment and the design of the fixed installation is not affected because the circuits in general still have to be designed, as indicated in *Regulations 471-17* and *471-18*, to be suitable also for the connection of Class I equipment and all socket outlet circuits must therefore incorporate provision for a protective conductor.

The second part of *Regulation 471-17* is intended to cover cases such as that of a Class II luminaire having external metal parts, where that luminaire is fixed into a metal BS box which in turn is mounted on, for instance, a metal stanchion or cable trunking connected either directly or indirectly to the main earthing terminal of the installation.

Where this measure is used for fixed current-using equipment, the responsibility of the designer is usually limited to selecting items of equipment of Class II construction complying with an appropriate British Standard. Again the choice of such equipment has little effect on the design of circuits which must still be designed to incorporate a protective conductor because of the possibility that the equipment may later be replaced by Class I equipment.

Item (i) of *Regulation 413-18* refers specifically to factory-built assemblies to BS 5486 and requires that they shall have, what is termed in that standard, 'total insulation'. This, in fact, is what is called in other equipment standards 'insulation encased Class II construction' to which reference has already been made here, and *Regulations 413-19* to *413-25* follow very closely the requirements in BS 5486.

In special cases use can be made of Items (ii) and (iii) of *Regulation 413-18* which allow supplementary or reinforced insulation to be applied on site during erection of the installation, but where this is done it has to be verified that the final result gives the same degree of safety as factory-built Class II equipment. To check compliance would involve the use of special test gear including a high-voltage tester and, for instance, the application of metal foil over the insulating enclosure to which the test voltage is applied, but as such tests are generally inconvenient to make on site the use of factory-built Class II equipment is always to be preferred.

### 4.6.3 Protection by non-conducting location

The next protective measure against indirect contact is termed 'protection by non-conducting location' and has a very limited application. As indicated in *Regulation 471-19*, it may only be used in special situations under effective supervision and where specified by a suitably qualified engineer. *Regulation 413-27* requires that steps must be taken to ensure that there is adequate physical separation between exposed conductive parts and extraneous conductive parts which could be at different potentials if there was a failure of the basic insulation of the equipment concerned. Furthermore, *Regulations 413-28* and *413-29* indicate that the second line of defence against indirect contact is obtained by the nature of the location itself which is such that the walls and floor of that location do not constitute a path for shock currents.

The measure should not be confused with the use of a non-conducting location intended to give protection against direct contact in situations where work is to be carried out by skilled persons on, or in the vicinity of, live parts, although the requirements for both cases are very similar.

There are very few locations where the continuing effectiveness of this measure can be guaranteed and the Note to *Regulation 413-30* indicates two possible ways in which the measure could be invalidated. It is also difficult to ensure compliance with *Regulation 413-31* because, for instance, a fault voltage appearing on conductive parts within the non-conducting location could be propagated outside the location through extraneous parts such as water pipes which had been added after completion of the installation. Hence the need for effective supervision.

### 4.6.4 Protection by earth free local equipotential bonding

The protective measure 'protection by earth free local equipotential bonding' also has a limited application. It can only be used under effective supervision and where specified by a suitably qualified electrical engineer. The correct implementation of the requirements of *Regulations 413-32* to *413-34* results in a true 'Faraday Cage' and prevents the appearance of any dangerous voltage between simultaneously accessible parts within the location concerned.

This measure cannot be applied to an entire installation and it is difficult to coordinate safely with other protective measures used elsewhere in the installation. In particular, *Regulation 413-34* indicates the danger which can exist at the point of transition between the location where the measure is applied and other locations.

### 4.6.5 Protection by electrical separation

The last protective measure against indirect contact in Section 413 is termed 'protection by electrical separation'. This measure is for use in an individual circuit and preferably that circuit should be intended to supply only one item of electrical equipment. As indicated by *Regulation 413-36*, the source of supply to the circuit has to have a high degree of isolation from its primary side and be of Class II construction or the equivalent. This regulation also indicates that there is to be no connection at the source to Earth so that no path for shock currents to Earth exists in the event of failure of basic insulation.

Bearing in mind that the nominal voltage of a circuit using this protective measure is permitted by *Regulation 413-36* (iv) to be as high as 500 V it is also important to ensure that this circuit - (the 'separated circuit') - is not accidentally earthed, because should this happen the circuit and its supply become a TN system and the person protected would be at risk from the voltages which could then appear between simultaneously accessible parts in the event of a second earth fault; the circuit has also to be protected against overcurrent. It becomes increasingly difficult to guarantee that the circuit will not become accidentally earthed as its length increases and this is particularly so if the circuit includes flexible cables which may be susceptible to mechanical damage, hence item (ii) of *Regulation 413-37* which requires flexible cables and cords to be visible where they are liable to such damage.

Item (iii) of *Regulation 413-37* permits the separated circuit to be in a multicore cable with other circuits or to be accommodated in the same conduit as other cables, provided that neither the sheath of that cable nor the conduit is metallic, again because of the risk that the separated circuit could become accidentally earthed via such a sheath or conduit.

With regard to item (iii) of *Regulation 413-37*, it must be pointed out that any separated circuit has to be protected against overcurrent unless exempted by Chapter 43 of the Wiring Regulations but item (iii) is intended to show that when the separated circuit shares the same cable or conduit with another circuit (which may be another separated circuit or a 'normal' circuit) no exemption from the provision of overcurrent protection for each circuit is permitted.

*Regulation 413-39* item (i) requires that where the separated circuit feeds more than one item of current-using equipment the exposed metalwork of those items should be connected together but not to such metalwork of the source. A first fault to the metalwork of one of the items of equipment presents neither a shock nor fire hazard but if another fault occurs in the same equipment, fed by the conductor of different polarity, this is effectively a short circuit and although there is still no shock hazard the overcurrent protective device must operate sufficiently rapidly to disconnect the circuit before the conductors reach an excessive temperature. If the second fault occurred in another item of equipment, again fed by the conductor of different polarity from that associated with the first fault, then *without* the bonding conductor connecting the exposed metalwork of the two equipments concerned there could be full voltage between the exposed metalwork thereby creating a shock hazard. As a further precaution, it is required by item (iv) of *Regulation 413-39* that the overcurrent protective device of the circuit clears the fault within the time specified in *Regulation 413-4* appropriate to the type of equipment concerned.

The measure should not be confused with the use of what is sometimes termed electrical separation, generally a safety isolating transformer, as a means of protection against electric shock in situations where work is to be done on or in the vicinity of live parts.

### 4.6.6 Reduction in nominal voltages as a protective measure

The Wiring Regulations recognise three types of system having reduced nominal voltages, these being
(i) the reduced low voltage system (for which the requirements are given in *Regulation 471-28* to *471-33*)
(ii) safety extra-low voltage systems (for which the requirements are given in *Regulations 411-2* to *411-10* and *Regulations 471-2* and *471-3*)
(iii) functional extra-low voltage systems (for which the requirements are given in *Regulations 411-11* to *411-15* and *Regulation 471-4*).

### 4.6.7 Reduced low voltage systems

The first of these systems, as stated in *Regulation 471-28*, has a nominal voltage not exceeding 110 V (r.m.s. a.c.) between phases, and is normally derived from a transformer or motor-generator set which, if single-phase, has the midpoint of its output winding directly earthed so that the voltage to earth ($U_0$) is half the phase voltage or, if three-phase, has the neutral point of that winding earthed so that $U_0$ is 0·577 times the line-to-line voltage.

Probably the most common form which is encountered is what is usually called the '110 volt centre tap earthed system' that was primarily developed for use on building and construction sites and other applications involving the large scale use of portable electric tools. This system was embodied in the British Standard dealing with the distribution units for these sites (BS 4363) and the companion Code of Practice (CP 1017) published in 1968 and 1969, respectively, although, even some twenty years earlier, in the 1949 Annual Report of HM Chief Inspector of Factories, there was a strong recommendation to adopt the system for those applications.

*Regulation 471-32* requires that protection against indirect contact has to be provided by means of an overcurrent protective device in each phase conductor or by a residual current device and BS 4363 requires the former to be a miniature circuit breaker. When the nominal voltage is 110 V and an overcurrent protective device is used for compliance with *Regulation 413-4*, the earth fault loop impedance of any circuit must not exceed the appropriate value in Table 41A2 (because the disconnection time must not exceed 5 s) multiplied by 0·23, (i.e. 55/240) if single phase, and by 0·26 (i.e. 63·5/240) if three-phase.

CP 1017 recommends that site lighting other than floodlighting should also be supplied from a 110 V reduced voltage source as should portable handlamps for general use. For portable handlamps used in confined or damp situations the Code of Practice recommends 50 V and 25 V as the nominal voltage.

### 4.6.8 Safety extra-low voltage systems

When considering the two extra-low voltage systems, it should be borne in mind that it has been accepted by the international committees concerned with installation rules, in both IEC and CENELEC, that in normally dry situations, i.e. where the person protected has a conventionally normal body resistance as mentioned in the Note to *Regulation 471-2* and is not in direct contact with Earth, a voltage of less than 50 V r.m.s. a.c. or 120 V ripple-free d.c. appearing across simultaneously accessible parts in the event of an earth fault can be tolerated indefinitely. If that voltage is 50 V a.c. (or 120 V d.c.) the fault is normally required to be cleared within 5 s.

Thus both the safety extra-low voltage and functional extra-low voltage systems are limited to a nominal voltage of 50 V r.m.s. a.c. or 120 V ripple-free d.c., but lower values than these maxima have to be used in other than normally dry situations, as indicated in *Regulation 471-2*. For instance, in agricultural installations, *Regulation 471-41* requires that where a safety extra-low voltage system is used, the nominal voltage must be less than 50 V a.c. (or 120 V d.c.) and, as suggested in the Note to *Regulation 471-40* this voltage should, in fact, be less than 25 V a.c., (or 60 V d.c.).

Examination of *Regulations 411-3* to *411-9* will immediately indicate that there are two essential features of a safety extra-low voltage system. First, no point in such a system or circuit may be earthed either directly, or indirectly through connection to protective conductors or exposed conductive parts of another system or through extraneous conductive parts. Secondly, there must be a high degree of electrical separation from other systems by using a good quality safety source and by arranging, if possible, that the safety extra-low voltage circuit is physically separated from other circuits or, if not, that the safety extra-low voltage circuit is adequately insulated from other circuits in order to ensure that no dangerous potential can be imposed on that circuit from any other source, even under fault conditions.

As a general indication of the degree of electrical separation for a safety extra-low voltage circuit as required by *Regulation 411-6*, Table 2 indicates the minimum creepage distances and clearances, and distances through insulation which should be adopted. In particular, the tabulated distances and clearances should be provided between the live parts of equipment such as relays, contactors, auxiliary switches and any part of any other circuit. It is also pointed out that the tabulated values assume clean, dry and vibration-free conditions only and should be appropriately increased if other environmental conditions are expected.

In addition, the arrangement used to attain the specified electrical separation shall withstand, for one minute, an applied voltage of 4 kV when the nominal voltage of the higher voltage circuit is 380 V or less, or 6 kV when the nominal voltage of the higher voltage circuit exceeds 380 V (but is less than 1000 V), and this requirement also follows BS 3535.

Should a fault develop in a safety extra-low voltage circuit, for instance, to an earthed extraneous conductive part, the person protected will not experience an electric shock unless a second fault occurs. If that should happen the maximum voltage of the circuit and the system may be said to give intrinsic protection against indirect contact.

**Table 2: Creepage distances and clearances, and distances through insulation for safety extra-low voltage systems**

| | Nominal voltage of circuit from which the SELV equipment is to be separated | |
|---|---|---|
| | Up to and including 250 V | Above 250 V |
| *Creepage distances and clearances* | mm | mm |
| 1 (*a*) Between terminals other than those covered by Item 1b below | 10 | 25 |
| (*b*) Between any terminal of the SELV circuit and any terminal of the other circuit. | 25 | 25 |
| 2 Between live parts other than terminals: | | |
| (*a*) All bare live parts other than those covered by Item 2b below | 3 | 5 |
| (*b*) Bare live parts of SELV circuits and other circuits | 10 | 12 |
| 3 Between live parts and other metal parts: | | |
| (*a*) for basic insulation | 4 | 5 |
| (*b*) for reinforced insulation. | 8 | 12 |
| 4 Between metal parts separated by basic insulation | 4 | 6 |
| *Distances through insulation* | | |
| 5 Between metal parts separated by supplementary insulation. | 1 | 1 |
| 6 Between metal parts separated by reinforced insulation. | 2 | 2 |

NOTES 1: The values given against item 1 do not apply to contact clearances of switches.

Table 2 has been based on the corresponding table in BS 3535.

There is therefore no need to have any other form of protection against indirect contact for a safety extra-low voltage circuit. The conductors of that circuit, however, must be protected against overcurrent for compliance with Chapter 43 of the Wiring Regulations.

Where an extra-low voltage circuit is supplied from a higher voltage system by means such as autotransformers, potential dividers, and semiconductor devices, it cannot be a safety extra-low voltage circuit and must be regarded as an extension of the input circuit. It then has to incorporate protection appropriate to the type of input circuit.

If the installation is required to comply with the Electricity (Factories Act) Special Regulations, 1908 and 1944, the safety extra-low voltage measure is not at present suitable unless any portable current-using equipment being supplied is of Class II construction (double insulated, or the equivalent). The reason for this is mainly legal, because Regulation 13 of those Regulations requires exposed metalwork or portable equipment to be earthed, and the only exemption to this so far allowed (in 1968) relates to Class II equipment.

### 4.6.9 Functional extra-low voltage systems

The third and last system to be considered here is that termed 'functional extra-low voltage' which, like safety extra-low voltage, has a maximum nominal voltage of 50 V r.m.s. a.c. or 120 V d.c., but as indicated in *Regulation 471-4* differs from the latter system in certain respects. The nominal voltage has to be less than the above values where other than normal dry conditions are likely to occur.

Functional extra-low voltage is usually selected simply because it suits the needs of the equipment to be supplied and the Note to *Regulation 411-11* gives an example of one way in which a functional extra-low voltage circuit can differ from safety extra-low voltage, namely, by being earthed at some point. In such a case, as indicated by *Regulation 411-12*, no other protection is required against indirect contact because the source is a safety source and it can be assumed that the maximum voltage attained under fault conditions will not exceed the nominal value of functional extra-low voltage concerned, and neither will a voltage from the primary side of that source be impressed on the functional extra-low voltage circuit.

*Regulation 411-13* deals with the case where a functional extra-low voltage system departs in more than one aspect from the safety extra-low voltage system, in particular, that the source is not a safety source, so that it must now be assumed that there is a possibility of a voltage from the primary side of the source being impressed on the functional extra-low voltage circuit. Here the exposed conductive parts of the extra-low voltage circuit, in accordance with *Regulation 411-14*, have to be connected to the protective conductor of the primary circuit.

## 4.7 Protection against both direct and indirect contact

The two extra-low voltage systems just considered appear in the Wiring Regulations under the above heading, i.e. in Section 411. It has been seen that in the case of safety extra-low voltage systems protection against indirect contact is 'intrinsic', whereas this is true of a functional extra-low voltage circuit only if it is fed from a safety source.

*Regulation 411-10* indicates that only if the nominal voltage of a safety extra-low voltage system does not exceed 25 V r.m.s. a.c. or 60 V ripple-free d.c. can it be claimed to meet the requirements for protection against *direct* contact without taking any further steps such as the adoption of one of the measures detailed in Section 412 of the Wiring Regulations and considered earlier in this present chapter. In other words it is considered safe, *in normal dry conditions*, to touch bare live parts at those voltages and only in the voltage range 25 V to 50 V a.c. or 60 V to 120 V d.c. is it necessary to adopt one of the measures in Section 412 (e.g. insulation of live parts) in order to give protection against direct contact. Where a functional extra-low voltage system is used and has a safety source, *Regulation 411-12* indicates the two protective measures against direct contact which must be used. If that system differs from safety extra-low voltage in more than one aspect including having a source which is not a safety source, *Regulation 411-13* indicates the appropriate protective measures then to be used for protection against direct contact.

The only other protective measure recognised by the Fifteenth Edition as giving protection against both direct and indirect contact is termed 'protection by limitation of discharge of energy', but no detailed requirements are given in *Regulation 411-16* other than requiring circuits in which this measure is used to be separated from other circuits in the same manner as that prescribed for safety extra-low voltage circuits. *Regulation 471-5* indicates that, in general, this protective measure is applied only to individual items of current-using equipment but there may be cases where the application of the measure can be extended to the part of the installation derived from such equipment. The measure is attained in mains operated electric fence controllers to BS 2632 by limiting the magnitude, duration and frequency of output pulses and of the energy and current that can flow during a specified period.

Another method is by incorporation in the equipment of a high output impedance (protective impedance) such that only a limited current can flow, such as for external connections to television receivers complying with BS 415.

Intrinsically safe circuits and equipment used in potentially explosive atmospheres are yet another example of a protective measure affording protection against both direct and indirect contact.

## 4.8 Earthing arrangements

In Chapter 3 of this Commentary, mention has already been made of the fact that the supply undertakings have the declared aim of being able to offer to all consumers an earthing terminal connected by a continuous metallic path to the earthed point of the source of energy, i.e. to the neutral points of the supply undertakings' distribution transformers. *Regulations 542-2* to *542-5* merely indicate the manner in which this connection is made at the consumer's installation and require no comment here.

With regard to *Regulation 542-7,* the actual resistance value from the consumer's earthing terminal to the earthed point of the source of energy is, for an installation which is part of one of the TN family of systems, outside the control of the installation designer if the installation is fed directly from the public supply network. This is only partly true if the installation is part of a TT system because the installation earth electrode, the resistance of which may be the dominant part of the earth fault loop impedance, is now the responsibility of the installation designer. In this latter case the designer still has no control over the resistance value of the earth electrode at the supply undertaking's source of energy.

When an installation earth electrode is necessary, *Regulations 542-10* to *542-15* are required to be met and the designer will find useful information to assist him in CP 1013 'Earthing'. What is important to remember is that not only should an earth electrode be satisfactory when first installed but that appropriate measures are taken to see that it continues to be so.

## 4.9 Protective conductors

The terms used to describe the various types of protective conductor and where each type appears in an installation have been commented on earlier in this present chapter. Here the aim is to indicate how a protective conductor in addition to the limitations placed on it in respect of the contribution it makes to the earth fault loop impedance has also to meet the requirements of Chapter 54 of the Wiring Regulations in order that it does not reach a dangerous temperature during an earth fault. For some types of protective conductor there are other limits of minimum cross-sectional area imposed by considerations of resistance to corrosion and to mechanical damage.

If a protective conductor serves more than one of the functions of earthing conductor, circuit protective conductor, or bonding conductor, it has to satisfy all the requirements for the functions involved. The size of the protective conductor will therefore be determined by the most onerous of the requirements for the particular functions concerned.

*Regulations 542-16* to *542-18* concern earthing conductors and the first of these regulations indicates that these conductors must comply with Section 543. This in turn allows the designer to arbitrarily select the cross-sectional areas from Table 54F or to calculate the minimum cross-sectional areas using the equation given in *Regulation 543-2*. In order to be able to adopt the latter alternative, it is necessary for the designer to know the magnitude of the fault current which can flow and its duration, and for these earthing conductors the fault current is limited only by that part of the earth fault loop impedance which is external to the installation (one of the characteristics of the supply which has to be ascertained as indicated in *Regulation 313-1*). When the installation is to be supplied from the public supply network, it may be necessary to obtain guidance in this respect from the supply undertaking and it will also be necessary to have information on the time/current characteristic of the supply undertaking's service fuses. In the absence of the necessary information from the supply undertaking there is no alternative but to use Table 54F.

For an earthing conductor buried in the ground, minimum cross-sectional areas are specified in Table 54A which take account of resistance to mechanical damage and/or corrosion and it should be noted that the cross-sectional area finally chosen for an earthing conductor will automatically give that for the main equipotential bonding conductor, as indicated by *Regulation 547-2*.

It will be noted from Table 18 of this Commentary (in Chapter 12) that because the size of the main equipotential bonding conductor is directly related, by *Regulation 547-2*, to the size of the earthing conductor, the former can in some cases be unduly large compared with the phase conductor sizes in the installation. If a change in the cross-sectional area of a buried earthing conductor is made where it comes out of the ground the size of the main equipotential bonding conductor can then be related to the reduced earthing conductor size.

For circuit protective conductors, the designer again has the choice of either arbitrarily choosing the cross-sectional area from application of Table 54F or

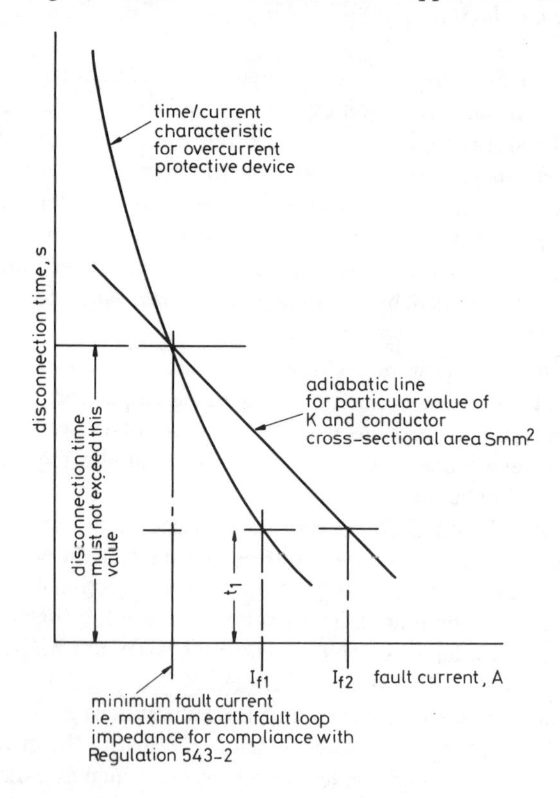


**Fig. 10** *Graphical determination of earth fault loop impedance for compliance with Regulation 543-2*

If the fault current is $I_{f1}$ A, the corresponding disconnection time is $t_1$ s, and as this current is less than $I_{f2}$, which is the maximum current which can be tolerated by the conductor concerned for that time, *Regulation 543-2* is met.

calculating that area using the equation and appropriate $k$ values given in *Regulation 543-2*. Again, as indicated in *Regulation 543-1*, there is an absolute minimum cross-sectional area determined by consideration of mechanical damage and corrosion if the protective conductor concerned is 'separate' as defined in the regulation.

The equation in *Regulation 543-2* is exactly the same as that given in *Regulation 434-6* although expressed differently and its derivation is quoted in Chapter 6 of this Commentary. It is called the adiabatic equation because it is based on the assumption that although the earth fault current may be many times the normal current-carrying capacity of the conductor concerned, the time during which that current is sustained is so short that all the heat energy generated is retained in the conductor.

Strictly, the equation is applicable only to cable conductors of cross-sectional area of 10 $mm^2$ or greater (see *Regulation 434-6*) and there is some investigational work being undertaken on smaller conductor sizes. Provided that the disconnection time achieved is less than 0·5 s or thereabouts, it is believed that the values of $k$ given in the regulation apply with no significant loss of accuracy to conductor cross-sectional areas down to 1 $mm^2$ and for longer disconnection times the application of the equation and the given $k$ values errs on the safe side.

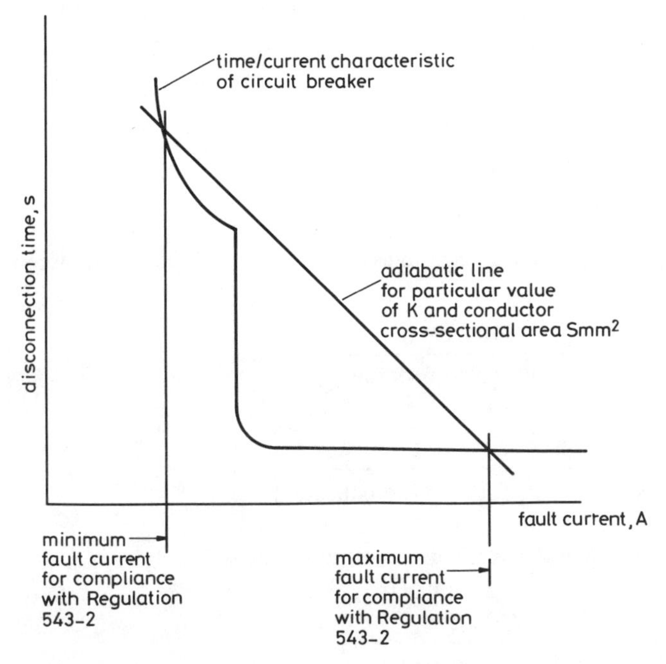


**Fig. 11** *Limitation of earth fault current when overcurrent protective device is a circuit breaker*

Time/current characteristics are usually plotted on double logarithmic graph paper and the adiabatic equation on such paper is a straight line. As shown in Fig. 10, one can very readily establish graphically the minimum earth fault current for compliance with *Regulation 543-2* using the time/current characteristic for the device concerned and the adiabatic line appropriate to the cross-sectional area and $k$ factor of the protective conductor it is intended to use.

For miniature circuit breakers and moulded-case circuit breakers, there is a maximum fault current in addition to the minimum value which can be tolerated thermally, as shown in Fig. 11. This Figure applies equally to short circuit conditions.

Where protection against indirect contact is afforded by overcurrent protective devices, the values of maximum earth fault impedances for disconnecting times not exceeding 0·4 s and 5 s are given in Tables 41A1 and 41A2, respectively, in the Wiring Regulations, and these, associated with the equation in *Regulation 543-2*, will give the corresponding minimum cross-sectional areas of protective conductor one can use.

The equation in *Regulation 543-2* can be written

$$S = \frac{I\sqrt{t}}{k} \text{mm}^2$$

and as

$$I = \frac{U_0}{Z_s} \text{ amperes}$$

$$S = \frac{U_0\sqrt{t}}{Z_s k} \text{ mm}^2$$

When $U_0$, the nominal voltage to earth, is 240 V and the actual disconnection time is 0·4 s

$$S = \frac{240\sqrt{0{\cdot}4}}{Z_s \text{ for } 0{\cdot}4 \text{ s disconnection}} \frac{1}{k} \text{ mm}^2$$

$$= \frac{151{\cdot}8}{k} \frac{1}{Z_s \text{ for } 0{\cdot}4 \text{ s disconnection}} \text{ mm}^2$$

and similarly when the actual disconnection time is 5 s.

$$S = \frac{240\sqrt{5}}{Z_s \text{ for 5 s disconnection}} \frac{1}{k} \text{ mm}^2$$

$$= \frac{536{\cdot}7}{k} \frac{1}{Z_s \text{ for 5 s disconnection}} \text{ mm}^2$$

Using these two equations, Tables 3 to 6 have been produced for every value of $k$ quoted in Tables 54B, 54C and 54D of the Wiring Regulations and give the minimum cross-sectional areas of protective conductors. As demanded by *Regulation 543-2*, where the formula has resulted in a nonstandard size the nearest larger standard cross-sectional area to the tabulated values must be used.

Table 7 merely brings together the $k$ factors from Tables 54B, 54C and 54D of the Wiring Regulations. The $k$ values 115 and 143 are of particular general interest because the lower value relates to a copper conductor incorporated in a p.v.c.-insulated cable with the associated phase and neutral conductors while the upper value relates for instance to a separate p.v.c.-insulated single-core cable with copper conductor used as a protective conductor. It is for this reason that in Tables 3 to 6 the entries concerning these two $k$ values have been accentuated.

The value taken for $U_0$, i.e. 240 V, is merely a convenient way of using Tables 41A1 and 41A2 in order to obtain the fault current corresponding to the disconnection times of 0·4 s and 5 s. The Tables, in fact, are independent of the value of $U_0$.

Only one Table is required for miniature circuit breakers because, as already indicated, the limiting values of earth fault loop impedance are based on the instantaneous tripping currents of such devices which corresponds to a disconnection time not exceeding 0·1 s.

In this case,

$$S = \frac{a\, I_n \sqrt{0{\cdot}1}}{k} \text{ mm}^2$$

$a$ being a multiple of $I_n$, the rated current of the miniature circuit breaker, i.e. $aI_n$ is the instantaneous tripping current.

For Type 1 m.c.b.s, $a = 4$
For Type 2 m.c.b.s, $a = 7$
For Type 3 m.c.b.s, $a = 10$

These Tables represent an alternative design method because having determined the $k$ value appropriate to the material and type of circuit protective conductor from Tables 54B, 54C or 54D and making the assumption that the disconnection time will be the maximum permitted by Chapter 41, the minimum cross-sectional area of the circuit protective conductor is immediately obtained from the Table corresponding to the current rating of the overcurrent protective device concerned. It then remains to be checked that the earth fault loop impedance of the circuit is such that the specified disconnection time is not exceeded.

It is emphasised that the foregoing tables merely give the calculated values of the minimum cross-sectional areas of protective conductors for compliance with *Regulation 543-2* based on the assumption that the *actual* disconnection times are equal to the maximum time specified, i.e. either 0·4 s or 5 s. There may well be other design aspects which require one to use a larger size than that tabulated. For instance if aluminium conductors are used, *Regulation 521-1* requires them to

**Table 3: Minimum cross-sectional areas (in mm²) of protective conductors for compliance with *Regulation 543-2* assuming actual disconnection time is 0.4 s**

| k | Type and rating of overcurrent protective device | | | | | | | | | | | | | | | | | |
|---|---|---|---|---|---|---|---|---|---|---|---|---|---|---|---|---|---|---|
| | Fuses to BS 3036 | | | | | Fuses to BS 88 Pt. 2 | | | | | | | Fuses to BS 1361 | | | | | Fuse to BS 1362 |
| | 5A | 15A | 20A | 30A | 45A | 6A | 10A | 16A | 20A | 25A | 32A | 40A | 5A | 15A | 20A | 30A | 45A | 13A |
| 22 | 0.72 | 2.55 | 3.83 | 6.27 | 11.5 | 0.79 | 1.3 | 2.46 | 3.83 | 4.60 | 6.27 | 8.63 | 0.61 | 2.03 | 3.83 | 5.75 | 11.5 | 2.76 |
| 26 | 0.61 | 2.16 | 3.24 | 5.31 | 9.73 | 0.67 | 1.1 | 2.09 | 3.24 | 3.90 | 5.31 | 7.30 | 0.51 | 1.72 | 3.24 | 4.87 | 9.73 | 2.34 |
| 27 | 0.59 | 2.08 | 3.12 | 5.11 | 9.37 | 0.65 | 1.06 | 2.01 | 3.12 | 3.75 | 5.11 | 7.03 | 0.49 | 1.65 | 3.12 | 4.69 | 9.37 | 2.25 |
| 44 | 0.40 | 1.28 | 1.92 | 3.14 | 5.75 | 0.40 | 0.65 | 1.23 | 1.92 | 2.30 | 3.14 | 4.31 | 0.30 | 1.01 | 1.92 | 2.87 | 5.75 | 1.38 |
| 51 | 0.31 | 1.10 | 1.66 | 2.71 | 4.97 | 0.34 | 0.56 | 1.06 | 1.66 | 2.00 | 2.71 | 3.73 | 0.26 | 0.88 | 1.66 | 2.48 | 4.97 | 1.19 |
| 52 | 0.30 | 1.08 | 1.62 | 2.65 | 4.87 | 0.33 | 0.55 | 1.04 | 1.62 | 1.95 | 2.65 | 3.65 | 0.26 | 0.86 | 1.62 | 2.43 | 4.87 | 1.17 |
| 54 | 0.29 | 1.04 | 1.56 | 2.55 | 4.68 | 0.32 | 0.53 | 1.00 | 1.56 | 1.87 | 2.55 | 3.51 | 0.25 | 0.83 | 1.56 | 2.34 | 4.68 | 1.12 |
| 60 | 0.26 | 0.94 | 1.41 | 2.30 | 4.22 | 0.29 | 0.48 | 0.90 | 1.41 | 1.69 | 2.30 | 3.16 | 0.22 | 0.74 | 1.41 | 2.11 | 4.22 | 1.01 |
| 64 | 0.25 | 0.88 | 1.32 | 2.15 | 3.95 | 0.27 | 0.45 | 0.85 | 1.32 | 1.58 | 2.15 | 2.96 | 0.21 | 0.70 | 1.32 | 1.97 | 3.95 | 0.95 |
| 76 | 0.21 | 0.74 | 1.11 | 1.82 | 3.33 | 0.23 | 0.38 | 0.71 | 1.11 | 1.33 | 1.82 | 2.50 | 0.18 | 0.59 | 1.11 | 1.67 | 3.33 | 0.80 |
| 81 | 0.20 | 0.69 | 1.04 | 1.70 | 3.12 | 0.22 | 0.35 | 0.67 | 1.04 | 1.25 | 1.70 | 2.34 | 0.16 | 0.55 | 1.04 | 1.56 | 3.12 | 0.75 |
| 89 | 0.18 | 0.63 | 0.95 | 1.55 | 2.85 | 0.20 | 0.32 | 0.61 | 0.95 | 1.14 | 1.55 | 2.14 | 0.15 | 0.50 | 0.95 | 1.42 | 2.85 | 0.68 |
| 93 | 0.17 | 0.60 | 0.91 | 1.48 | 2.72 | 0.19 | 0.31 | 0.58 | 0.91 | 1.09 | 1.48 | 2.04 | 0.14 | 0.48 | 0.91 | 1.36 | 2.72 | 0.65 |
| 94 | 0.17 | 0.60 | 0.90 | 1.47 | 2.69 | 0.19 | 0.30 | 0.58 | 0.90 | 1.08 | 1.47 | 2.02 | 0.14 | 0.47 | 0.90 | 1.35 | 2.69 | 0.65 |
| 95 | 0.17 | 0.59 | 0.89 | 1.45 | 2.67 | 0.18 | 0.30 | 0.57 | 0.89 | 1.07 | 1.45 | 2.00 | 0.14 | 0.47 | 0.89 | 1.33 | 2.67 | 0.64 |
| 98 | 0.16 | 0.57 | 0.86 | 1.41 | 2.58 | 0.18 | 0.29 | 0.55 | 0.86 | 1.03 | 1.41 | 1.94 | 0.14 | 0.46 | 0.86 | 1.29 | 2.58 | 0.62 |
| 110 | 0.14 | 0.51 | 0.77 | 1.25 | 2.30 | 0.16 | 0.26 | 0.49 | 0.77 | 0.92 | 1.25 | 1.73 | 0.12 | 0.41 | 0.77 | 1.15 | 2.30 | 0.55 |
| **115** | **0.14** | **0.49** | **0.73** | **1.20** | **2.20** | **0.15** | **0.25** | **0.47** | **0.73** | **0.88** | **1.20** | **1.65** | **0.12** | **0.39** | **0.73** | **1.10** | **2.20** | **0.53** |
| 116 | 0.14 | 0.48 | 0.73 | 1.19 | 2.18 | 0.15 | 0.25 | 0.47 | 0.73 | 0.87 | 1.19 | 1.64 | 0.11 | 0.38 | 0.73 | 1.09 | 2.18 | 0.52 |
| 134 | 0.12 | 0.42 | 0.63 | 1.03 | 1.89 | 0.13 | 0.21 | 0.40 | 0.63 | 0.76 | 1.03 | 1.42 | 0.10 | 0.33 | 0.63 | 0.94 | 1.89 | 0.45 |
| **143** | **0.11** | **0.39** | **0.59** | **0.96** | **1.77** | **0.12** | **0.20** | **0.38** | **0.59** | **0.71** | **0.96** | **1.33** | **0.09** | **0.31** | **0.59** | **0.88** | **1.77** | **0.42** |
| 166 | 0.10 | 0.34 | 0.51 | 0.83 | 1.52 | 0.11 | 0.17 | 0.33 | 0.51 | 0.61 | 0.83 | 1.14 | 0.08 | 0.27 | 0.51 | 0.76 | 1.52 | 0.36 |
| 176 | 0.09 | 0.32 | 0.48 | 0.78 | 1.47 | 0.10 | 0.16 | 0 31 | 0.48 | 0.57 | 0.78 | 1.08 | 0.08 | 0.25 | 0.48 | 0.72 | 1.44 | 0.34 |

**Table 4: Minimum cross-sectional areas (in mm²) of protective conductors for compliance with *Regulation 543-2* assuming actual disconnection time is 5 s**

| $k$ | Type and rating of overcurrent protective device | | | | | | | | | | | | | | |
|---|---|---|---|---|---|---|---|---|---|---|---|---|---|---|---|
| | Fuses to BS 3036 | | | | | | | Fuses to BS 1361 | | | | | | | |
| | 5A | 15A | 20A | 30A | 45A | 60A | 100A | 5A | 15A | 20A | 30A | 45A | 60A | 80A | 100A |
| 22 | 1.22 | 4.36 | 6.10 | 7.63 | 15.3 | 20.3 | 44.4 | 1.44 | 4.60 | 8.41 | 12.2 | 24.4 | 40.7 | 50.8 | 87.1 |
| 26 | 1.03 | 3.69 | 5.16 | 6.45 | 12.9 | 17.2 | 37.5 | 1.21 | 3.89 | 7.12 | 10.3 | 20.6 | 34.4 | 43.0 | 73.7 |
| 27 | 1.00 | 3.55 | 4.97 | 6.21 | 12.4 | 16.6 | 36.1 | 1.17 | 3.75 | 6.86 | 9.94 | 19.9 | 33.1 | 41.4 | 71.0 |
| 44 | 0.61 | 2.18 | 3.05 | 3.81 | 7.63 | 10.2 | 22.2 | 0.72 | 2.30 | 4.21 | 6.10 | 12.2 | 20.3 | 25.4 | 43.6 |
| 51 | 0.53 | 1.88 | 2.63 | 3.29 | 6.58 | 8.77 | 19.1 | 0.62 | 1.98 | 3.63 | 5.26 | 10.5 | 17.5 | 21.9 | 37.6 |
| 52 | 0.52 | 1.84 | 2.58 | 3.23 | 6.45 | 8.60 | 18.8 | 0.61 | 1.95 | 3.56 | 5.16 | 10.3 | 17.2 | 21.5 | 36.9 |
| 54 | 0.50 | 1.77 | 2.49 | 3.11 | 6.21 | 8.28 | 18.1 | 0.58 | 1.88 | 3.43 | 4.97 | 9.94 | 16.6 | 20.7 | 35.5 |
| 60 | 0.45 | 1.60 | 2.24 | 2.80 | 5.60 | 7.46 | 16.3 | 0.53 | 1.69 | 3.09 | 4.48 | 8.95 | 14.9 | 18.6 | 31.9 |
| 64 | 0.42 | 1.50 | 2.10 | 2.62 | 5.24 | 7.00 | 15.3 | 0.49 | 1.58 | 2.89 | 4.20 | 8.39 | 14.0 | 17.5 | 29.9 |
| 76 | 0.35 | 1.26 | 1.77 | 2.21 | 4.41 | 5.88 | 12.8 | 0.42 | 1.33 | 2.43 | 3.53 | 7.06 | 11.8 | 14.7 | 25.2 |
| 81 | 0.33 | 1.18 | 1.66 | 2.07 | 4.14 | 5.52 | 12.1 | 0.39 | 1.25 | 2.29 | 3.32 | 6.63 | 11.0 | 13.8 | 23.7 |
| 89 | 0.30 | 1.08 | 1.51 | 1.88 | 3.77 | 5.02 | 11.0 | 0.35 | 1.14 | 2.08 | 3.02 | 6.03 | 10.0 | 12.6 | 21.5 |
| 93 | 0.29 | 1.03 | 1.44 | 1.80 | 3.61 | 4.81 | 10.5 | 0.34 | 1.09 | 1.99 | 2.89 | 5.77 | 9.62 | 12.0 | 20.6 |
| 94 | 0.29 | 1.02 | 1.43 | 1.78 | 3.57 | 4.75 | 10.4 | 0.34 | 1.08 | 1.97 | 2.86 | 5.71 | 9.52 | 11.9 | 20.4 |
| 95 | 0.28 | 1.01 | 1.41 | 1.77 | 3.53 | 4.71 | 10.3 | 0.33 | 1.07 | 1.95 | 2.83 | 5.65 | 9.42 | 11.8 | 20.2 |
| 98 | 0.27 | 0.98 | 1.37 | 1.71 | 3.43 | 4.57 | 10.0 | 0.32 | 1.03 | 1.89 | 2.74 | 5.48 | 9.13 | 11.4 | 19.6 |
| 110 | 0.24 | 0.87 | 1.22 | 1.53 | 3.05 | 4.07 | 8.88 | 0.29 | 0.92 | 1.68 | 2.44 | 4.88 | 8.13 | 10.2 | 17.4 |
| **115** | **0.23** | **0.83** | **1.17** | **1.46** | **2.92** | **3.89** | **8.50** | **0.27** | **0.88** | **1.61** | **2.34** | **4.67** | **7.78** | **9.73** | **16.7** |
| 116 | 0.23 | 0.83 | 1.16 | 1.45 | 2.89 | 3.86 | 8.42 | 0.27 | 0.87 | 1.60 | 2.32 | 4.63 | 7.72 | 9.64 | 16.5 |
| 134 | 0.20 | 0.72 | 1.00 | 1.25 | 2.51 | 3.34 | 7.29 | 0.24 | 0.76 | 1.38 | 2.01 | 4.01 | 6.68 | 8.35 | 14.3 |
| **143** | **0.19** | **0.67** | **0.94** | **1.17** | **2.34** | **3.12** | **6.82** | **0.22** | **0.71** | **1.29** | **1.88** | **3.75** | **6.25** | **7.81** | **13.4** |
| 166 | 0.16 | 0.58 | 0.81 | 1.01 | 2.02 | 2.69 | 5.87 | 0.19 | 0.61 | 1.11 | 1.62 | 3.23 | 5.38 | 6.73 | 11.5 |
| 176 | 0.15 | 0.54 | 0.76 | 0.95 | 1.91 | 2.54 | 5.55 | 0.18 | 0.58 | 1.05 | 1.53 | 3.05 | 5.08 | 6.35 | 10.9 |

**Table 5: Minimum cross-sectional areas (in mm²) of protective conductors for compliance with *Regulation 543-2* assuming actual disconnection time is 5s**

| *k* | Fuses to BS88 Pt.2 | | | | | | | | | | | | | | | | | | |
|---|---|---|---|---|---|---|---|---|---|---|---|---|---|---|---|---|---|---|---|
| | 6A | 10A | 16A | 20A | 25A | 32A | 40A | 50A | 63A | 80A | 100A | 125A | 160A | 200A | 250A | 315A | 400A | 500A | 630A | 800A |
| 22 | 1.88 | 3.17 | 5.55 | 8.13 | 10.2 | 13.6 | 17.4 | 22.2 | 28.4 | 40.7 | 53.8 | 71.8 | 90.4 | 128 | 153 | 222 | 254 | 375 | 452 | 717 |
| 26 | 1.59 | 2.68 | 4.69 | 6.88 | 8.60 | 11.5 | 14.7 | 18.8 | 24.0 | 34.4 | 45.9 | 60.7 | 76.4 | 109 | 129 | 188 | 215 | 318 | 382 | 607 |
| 27 | 1.53 | 2.58 | 4.52 | 6.63 | 8.23 | 11.1 | 14.2 | 18.1 | 23.1 | 33.1 | 44.2 | 58.5 | 73.6 | 105 | 124 | 181 | 207 | 306 | 368 | 585 |
| 44 | 0.94 | 1.58 | 2.77 | 4.07 | 5.08 | 6.78 | 8.71 | 11.1 | 14.2 | 20.3 | 27.1 | 35.9 | 45.2 | 64.2 | 76.0 | 111 | 127 | 188 | 226 | 359 |
| 51 | 0.81 | 1.37 | 2.39 | 3.51 | 4.38 | 5.84 | 7.51 | 9.56 | 12.2 | 17.5 | 23.4 | 30.9 | 39.0 | 55.4 | 66.0 | 95.6 | 110 | 162 | 195 | 309 |
| 52 | 0.79 | 1.34 | 2.35 | 3.44 | 4.30 | 5.73 | 7.37 | 9.38 | 12.0 | 17.2 | 22.9 | 30.4 | 38.2 | 54.3 | 64.5 | 93.8 | 107 | 159 | 191 | 304 |
| 54 | 0.76 | 1.29 | 2.26 | 3.31 | 4.14 | 5.52 | 7.10 | 9.04 | 11.6 | 16.6 | 22.1 | 29.2 | 36.8 | 52.3 | 62.1 | 90.4 | 104 | 153 | 184 | 292 |
| 60 | 0.69 | 1.16 | 2.03 | 2.98 | 3.73 | 4.97 | 6.39 | 8.14 | 10.4 | 14.9 | 19.9 | 26.3 | 33.1 | 47.1 | 55.9 | 81.4 | 93.2 | 138 | 166 | 263 |
| 64 | 0.65 | 1.09 | 1.91 | 2.80 | 3.50 | 4.66 | 6.00 | 7.63 | 9.76 | 14.0 | 18.6 | 24.7 | 31.1 | 44.2 | 52.4 | 76.3 | 87.4 | 129 | 155 | 247 |
| 76 | 0.54 | 0.92 | 1.60 | 2.35 | 2.94 | 3.92 | 5.04 | 6.42 | 8.21 | 11.8 | 15.7 | 20.8 | 26.1 | 37.2 | 44.1 | 64.2 | 73.5 | 109 | 131 | 208 |
| 81 | 0.51 | 0.86 | 1.51 | 2.21 | 2.76 | 3.68 | 4.74 | 6.03 | 7.71 | 11.0 | 14.7 | 19.5 | 24.6 | 34.9 | 41.4 | 60.3 | 69.1 | 102 | 123 | 195 |
| 89 | 0.46 | 0.78 | 1.37 | 2.01 | 2.51 | 3.35 | 4.31 | 5.48 | 7.01 | 10.0 | 13.4 | 17.7 | 22.3 | 31.7 | 37.7 | 54.8 | 62.8 | 92.8 | 112 | 177 |
| 93 | 0.44 | 0.75 | 1.31 | 1.92 | 2.40 | 3.21 | 4.12 | 5.25 | 6.71 | 9.62 | 12.8 | 17.0 | 21.4 | 30.4 | 36.1 | 52.5 | 60.1 | 88.8 | 107 | 170 |
| 94 | 0.44 | 0.74 | 1.30 | 1.90 | 2.38 | 3.17 | 4.08 | 5.19 | 6.64 | 9.52 | 12.7 | 16.8 | 21.1 | 30.1 | 35.7 | 51.9 | 59.5 | 87.8 | 106 | 168 |
| 95 | 0.43 | 0.73 | 1.28 | 1.88 | 2.35 | 3.14 | 4.04 | 5.14 | 6.57 | 9.42 | 12.5 | 16.6 | 20.9 | 29.7 | 35.3 | 51.4 | 58.9 | 86.9 | 105 | 166 |
| 98 | 0.42 | 0.71 | 1.26 | 1.83 | 2.28 | 3.04 | 3.91 | 4.98 | 6.37 | 9.13 | 12.2 | 16.1 | 20.3 | 28.8 | 34.3 | 49.8 | 57.1 | 84.3 | 101 | 161 |
| 110 | 0.38 | 0.63 | 1.11 | 1.63 | 2.03 | 2.71 | 3.49 | 4.44 | 5.67 | 8.13 | 10.8 | 14.4 | 18.1 | 25.7 | 30.5 | 44.4 | 50.8 | 75.1 | 90.4 | 144 |
| **115** | **0.36** | **0.61** | **1.06** | **1.56** | **1.95** | **2.59** | **3.34** | **4.25** | **5.43** | **7.78** | **10.4** | **13.7** | **17.3** | **24.6** | **29.2** | **42.5** | **48.6** | **71.8** | **86.5** | **137** |
| 116 | 0.36 | 0.60 | 1.05 | 1.54 | 1.93 | 2.57 | 3.31 | 4.21 | 5.38 | 7.72 | 10.3 | 13.6 | 17.1 | 24.4 | 28.9 | 42.1 | 48.2 | 71.2 | 85.7 | 136 |
| 134 | 0.31 | 0.52 | 0.91 | 1.34 | 1.67 | 2.23 | 2.86 | 3.65 | 4.66 | 6.68 | 8.91 | 11.8 | 14.9 | 21.1 | 25.1 | 36.5 | 41.8 | 61.7 | 74.2 | 118 |
| **143** | **0.29** | **0.49** | **0.85** | **1.25** | **1.56** | **2.08** | **2.68** | **3.41** | **4.36** | **6.25** | **8.33** | **11.0** | **13.9** | **19.7** | **23.4** | **34.1** | **39.1** | **57.7** | **69.4** | **110** |
| 166 | 0.25 | 0.42 | 0.73 | 1.08 | 1.35 | 1.79 | 2.31 | 2.94 | 3.76 | 5.38 | 7.18 | 9.50 | 12.0 | 17.0 | 20.2 | 29.4 | 33.6 | 49.7 | 59.8 | 95.0 |
| 176 | 0.23 | 0.40 | 0.69 | 1.02 | 1.27 | 1.69 | 2.18 | 2.77 | 3.55 | 5.08 | 6.78 | 8.97 | 11.3 | 16.1 | 19.1 | 27.7 | 31.8 | 46.9 | 56.5 | 90.0 |

**Table 6: Minimum cross-sectional areas (in mm²) of protective conductors for compliance with *Regulation 543-2* when using miniature circuit breakers and assuming actual disconnection time is 0.10 s**

| *k* | Type 1 | | | | | | Type 2 | | | | | | Type 3 | | | | | |
|---|---|---|---|---|---|---|---|---|---|---|---|---|---|---|---|---|---|---|
| | 5A | 10A | 15A | 20A | 30A | 50A | 5A | 10A | 15A | 20A | 30A | 50A | 5A | 10A | 15A | 20A | 30A | 50A |
| 22 | 0.29 | 0.58 | 0.86 | 1·15 | 1.73 | 2.88 | 0.50 | 1.01 | 1.51 | 2.01 | 3.02 | 5.03 | 0.72 | 1.44 | 2.16 | 2.87 | 4.31 | 7.19 |
| 26 | 0.24 | 0.49 | 0.73 | 0.97 | 1.46 | 2.43 | 0.43 | 0.85 | 1.28 | 1.70 | 2.55 | 4.26 | 0.61 | 1.22 | 1.82 | 2.43 | 3.65 | 6.08 |
| 27 | 0.23 | 0.47 | 0.70 | 0.94 | 1.41 | 2.34 | 0.41 | 0.82 | 1.23 | 1.64 | 2.46 | 4.10 | 0.59 | 1.17 | 1.76 | 2.34 | 3.51 | 5.86 |
| 44 | 0.14 | 0.29 | 0.43 | 0.58 | 0.86 | 1.44 | 0.25 | 0.50 | 0.75 | 1.01 | 1.51 | 2.52 | 0.36 | 0.72 | 1.08 | 1.44 | 2.16 | 3.60 |
| 51 | 0.12 | 0.25 | 0.37 | 0.50 | 0.74 | 1.24 | 0.22 | 0.43 | 0.65 | 0.87 | 1.30 | 2.17 | 0.31 | 0·62 | 0.93 | 1.24 | 1.86 | 3.10 |
| 52 | 0.12 | 0.24 | 0.36 | 0.49 | 0.73 | 1.22 | 0.21 | 0.43 | 0.64 | 0.85 | 1.28 | 2.13 | 0.30 | 0.61 | 0.91 | 1.22 | 1.82 | 3.04 |
| 54 | 0.12 | 0.23 | 0.35 | 0.47 | 0.70 | 1.17 | 0.21 | 0.41 | 0.62 | 0.82 | 1.23 | 2.05 | 0.29 | 0.59 | 0.88 | 1.17 | 1.76 | 2.93 |
| 60 | 0.11 | 0.21 | 0.32 | 0.42 | 0.63 | 1.05 | 0.18 | 0.37 | 0.55 | 0.74 | 1.11 | 1.85 | 0.26 | 0.53 | 0.79 | 1.05 | 1.58 | 2.64 |
| 64 | 0.10 | 0.20 | 0.30 | 0.40 | 0.59 | 0.99 | 0.17 | 0.35 | 0.52 | 0.69 | 1.04 | 1.73 | 0.25 | 0.49 | 0.74 | 1.00 | 1.48 | 2.47 |
| 76 | 0.083 | 0.17 | 0.25 | 0.33 | 0.50 | 0.83 | 0.15 | 0.29 | 0.44 | 0.58 | 0.87 | 1.46 | 0.21 | 0.42 | 0.62 | 0.83 | 1.25 | 2.08 |
| 81 | 0·078 | 0.16 | 0.23 | 0.31 | 0.47 | 0.78 | 0.14 | 0.27 | 0.41 | 0.55 | 0.82 | 1.37 | 0.20 | 0.39 | 0.59 | 0.78 | 1.17 | 1.95 |
| 89 | 0.071 | 0.14 | 0.21 | 0.28 | 0.43 | 0.71 | 0.12 | 0.25 | 0.37 | 0.50 | 0.75 | 1.25 | 0.18 | 0.36 | 0.53 | 0.71 | 1.07 | 1.78 |
| 93 | 0.068 | 0.14 | 0.20 | 0.27 | 0.41 | 0.68 | 0.12 | 0.24 | 0.36 | 0.48 | 0.71 | 1.19 | 0.17 | 0.34 | 0.51 | 0.68 | 1.02 | 1.70 |
| 94 | 0.067 | 0.13 | 0.20 | 0.27 | 0.40 | 0.67 | 0.12 | 0.24 | 0.35 | 0·47 | 0.71 | 1.18 | 0.17 | 0.34 | 0.50 | 0.67 | 1.01 | 1.68 |
| 95 | 0.067 | 0.13 | 0.20 | 0.27 | 0.40 | 0.67 | 0.12 | 0.23 | 0.35 | 0.47 | 0.70 | 1.17 | 0.17 | 0.33 | 0.50 | 0.67 | 1.00 | 1.67 |
| 98 | 0.065 | 0.13 | 0.19 | 0.26 | 0.39 | 0.65 | 0.11 | 0.23 | 0.34 | 0.45 | 0.68 | 1.13 | 0.16 | 0.32 | 0.48 | 0.65 | 0.97 | 1.62 |
| 110 | 0.058 | 0.12 | 0.17 | 0.23 | 0.35 | 0.58 | 0.10 | 0.20 | 0.30 | 0.40 | 0.60 | 1.01 | 0.14 | 0.29 | 0.43 | 0.57 | 0.86 | 1.44 |
| **115** | **0.055** | **0.11** | **0.17** | **0.22** | **0.33** | **0.55** | **0.10** | **0.19** | **0.29** | **0.38** | **0.58** | **0.96** | **0.14** | **0.28** | **0.41** | **0.55** | **0.83** | **1.38** |
| 116 | 0.055 | 0.11 | 0.16 | 0.22 | 0.33 | 0.55 | 0.10 | 0.19 | 0.29 | 0.38 | 0.57 | 0.96 | 0.14 | 0.27 | 0.41 | 0.55 | 0.82 | 1.37 |
| 134 | 0.047 | 0.094 | 0.14 | 0.19 | 0.28 | 0.47 | 0.08 | 0.17 | 0.25 | 0.33 | 0.50 | 0.83 | 0.12 | 0.24 | 0.35 | 0.47 | 0.71 | 1.18 |
| **143** | **0.044** | **0.089** | **0.13** | **0.18** | **0.27** | **0.44** | **0.08** | **0.16** | **0.23** | **0.31** | **0.47** | **0.78** | **0.11** | **0.22** | **0.33** | **0.44** | **0.66** | **1.11** |
| 166 | 0.038 | 0.076 | 0.11 | 0.15 | 0.23 | 0.38 | 0.07 | 0.13 | 0.20 | 0.27 | 0.40 | 0.67 | 0.10 | 0.19 | 0.29 | 0.38 | 0.57 | 0.96 |
| 176 | 0.036 | 0.072 | 0.11 | 0.14 | 0.22 | 0.36 | 0.06 | 0.13 | 0.19 | 0.25 | 0.38 | 0.63 | 0.09 | 0.18 | 0.27 | 0.36 | 0.54 | 0.90 |

have a cross-sectional area of at least 16 mm². Then again, if it is intended to use p.v.c.-insulated cables to BS 6004 the minimum cross-sectional area specified in that British Standard is 1 mm² which, in any event, is generally considered to be the practical minimum for fixed wiring from purely mechanical aspects of design and erection.

In some cases where the earth fault loop impedance is considerably less than that demanded by consideration of the specified maximum disconnection time (and this will frequently be so for the lower ratings of overcurrent protective devices), the actual cross-sectional area of protective conductor one can use will be less than the tabulated minimum because the earth fault current has been increased leading to a faster disconnection time.

This is best illustrated by Fig. 12.

This shows the time/current characteristic of the overcurrent protective device and the adiabatic lines for three cross-sectional areas of protective conductor $S_1$, $S_2$ and $S_3$ where $S_1 > S_2 > S_3$.

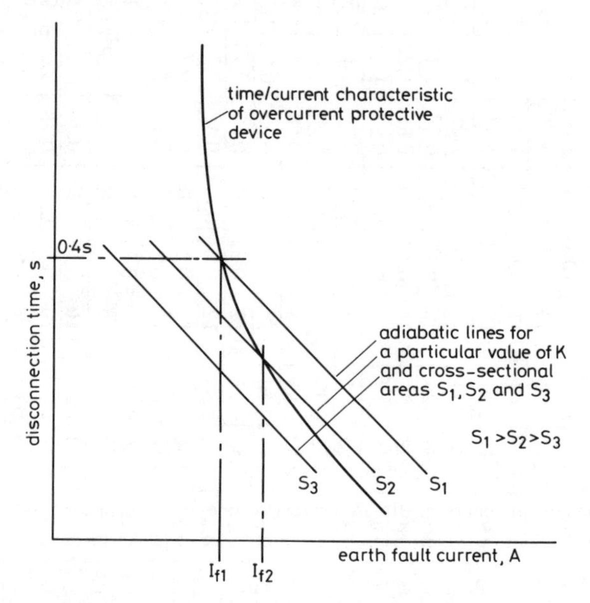


**Fig. 12** *Effect of change of cross-sectional area of protective conductor*
Both axes are logarithmic

$S_1$ is the tabulated value for the cross-sectional area corresponding to a disconnection time of 0·4 s so that the fault current is $I_{f1}$ amperes which in turn gives a maximum earth fault loop impedance $Z_{s1} = U_0/I_{f1}$ Ω. However, it is found that having designed the circuit to meet the voltage drop limitation of 2·5% and knowing the impedance of the earth loop external to the circuit, the actual earth fault loop impedance, even using a protective conductor area less than $S_1$, namely $S_2$, is still less than the maximum permissible for disconnection in 0·4 s.

The thermal requirements of *Regulation 543-2* are still met provided that this earth fault loop impedance is less than that given by $U_0/I_{f2}$ Ω where $I_{f2}$ is the current corresponding to the point of intersection of the time/current characteristic with the adiabatic line corresponding to $S_2$ mm².

For the same length of circuit, the situation could arise that one could use an even further reduced cross-sectional area of protective conductor, $S_3$ mm², from consideration only of the limitation of earth fault loop impedance. As shown in Fig. 12, the adiabatic line for this cross-sectional area lies completely below the time/current characteristic and *Regulation 543-2* would not be met. It will frequently happen that one intends to use a standard cable having a reduced protective conductor, e.g. one of the flat cables in BS 6004 and comparison of cross-sectional areas with the values in the appropriate table among the foregoing suggests that the protective conductor may not always be adequate. A practical example of this is as follows:

A single-phase 240 V circuit feeding fixed equipment is protected by a 45 A fuse to BS 1361. In order to comply with the voltage drop limitation of 2·5% under normal load conditions the resistance of the phase conductor must not exceed 0·067 Ω, assuming that the design current of the circuit is also 45 A. According to Table 9D2 of Appendix 9 of the Wiring Regulations, the circuit, if the installation method is chosen from methods E to H of Table 9A and the circuit is not grouped with other circuits, could be wired in 6 mm² 'twin and earth' flat p.v.c.-insulated cable to BS 6004 having a 2·5 mm² protective conductor. From Table 54C, the relevant value for $k$ is 115, and from the appropriate table amongst the foregoing the minimum cross-sectional area of the protective conductor should be 4·67 mm² and it therefore becomes necessary to check if it is possible to use 2·5 mm², bearing in mind that the disconnection time might be less than the permitted maximum of 5s.

For a 2·5 mm² protective conductor, its maximum resistance is (0·067 x 6)/2·5 Ω = 0·161 Ω, and if the impedance of that part of the earth fault loop external to the circuit is, for instance, known or assumed to be 0·5 Ω the earth fault loop impedance is (0·5 + 0·067 + 0·161) Ω i.e. 0·728 Ω, and as this is less than the 1·0 ohm maximum earth fault loop impedance for 5 s disconnection the next step is to calculate the actual disconnection time (which will be faster than 5 s).

The earth fault current is 240/0·728 A, i.e. 330 A, and from the time/current characteristic of the 45 A BS 1361 fuse it is found that the disconnection time is 1·2 s. It is now possible to check, using the adiabatic equation, whether 2·5 mm² cross-sectional area is adequate. But

$$S = \frac{I\sqrt{t}\ \mathrm{mm}^2}{k} = \frac{330\sqrt{1{\cdot}2}\ \mathrm{mm}^2}{115} = 3{\cdot}14\ \mathrm{mm}^2$$

Thus it will be seen that the 2·5 mm² protective conductor is not protected thermally if the earth fault loop impedance is 0·728 Ω and the earth fault current is 330 A.

**Table 7: *k*- factors for protective conductors**

| Description | Protective conductor material | Insulation of protective conductor or cable covering | | |
|---|---|---|---|---|
| | | P.V.C. | 90°C thermosetting | 85°C rubber |
| Insulated protective conductor not incorporated in cables, or bare protective conductor in contact with cable covering | copper | 143 | 176 | 166 |
| | aluminium | 95 | 116 | 110 |
| | steel | 52 | 64 | 60 |
| Protective conductor as a core in a cable | copper | 115 | 143 | 134 |
| | aluminium | 76 | 94 | 89 |
| Protective conductor as a sheath or armour of a cable | steel | 45 | 54 | 55 |
| | copper | – | – | – |
| | aluminium | 79 | 98 | 98 |
| | lead | 23 | 27 | 28 |

**Table 8: Gross cross-sectional areas of armour wires for multicore cables to BS 6346 having shaped conductors**

| Nominal cross-sectional area of conductor | Cross-sectional area of round armour wires | | | | | | | | |
|---|---|---|---|---|---|---|---|---|---|
| | Cables with stranded conductors | | | | | Cables with solid aluminium conductors | | | |
| | Two core | Three core | | Four core | | Two core | Three core | | Four core |
| | 600/1000V | 600/1000V | 1900/3300V | Equal neutral 600/1000V | Reduced neutral 600/1000V | 600/1000V | 600/1000V | 1900/3300V | 600/1000V |
| $mm^2$ | $mm^2$ | $mm^2$ | $mm^2$ | $mm^2$ | $mm^2$ | $mm^2$ | $mm^2$ | $mm^2$ | $mm^2$ |
| 16 | 39 | 43 | 76 | 64 | – | 35 | 41 | 72 | 60 |
| 25 | 61 | 67 | 84 | 76 | 76 | 55 | 64 | 80 | 71 |
| 35 | 66 | 75 | 91 | 85 | 82 | 60 | 70 | 86 | 79 |
| 50 | 76 | 85 | 124 | 123 | 94 | 68 | 80 | 117 | 114 |
| 70 | 85 | 122 | 138 | 139 | 135 | 75 | 113 | 129 | 129 |
| 95 | 123 | 140 | 152 | 160 | 158 | 110 | 130 | 142 | 148 |
| 120 | 133 | 151 | 207 | 221 | 220 | – | 140 | 193 | 203 |
| 150 | 146 | 190 | 221 | 243 | 235 | – | 196 | 205 | 224 |
| 185 | 205 | 234 | 238 | 269 | 262 | – | 216 | 221 | 248 |
| 240 | 230 | 262 | 262 | 308 | 293 | – | 243 | 243 | 279 |
| 300 | 254 | 289 | 289 | 335 | 324 | – | 268 | 268 | 308 |
| 400 | 281 | 322 | 322 | 474 | 459 | – | – | – | – |

Plotting the adiabatic line for 2·5 mm² over the time/current characteristic shows that the point of intersection corresponds to an earth fault current of 370 A, so in order to use the standard 6 mm² twin and earth cable with the 2·5 mm² protective conductor the earth fault loop impedance must not exceed 240/370 Ω, i.e. 0·65 Ω. This requirement can only be met if the impedance of that part of the earth fault loop external to the circuit is less than 0·5 Ω or the circuit length is reduced or the conductor size is increased.

In order to avoid the necessity of having to use the foregoing two-stage approach, for circuits where $U_0$ is 240 V, Appendix 8 of the Wiring Regulations includes Tables of limiting values of earth fault loop impedance when using p.v.c.-insulated cables to BS 6004, and if these values are met the circuit complies both with the appropriate disconnection time prescribed in *Regulation 413-4* and with the thermal limitation of *Regulation 543-2*. The above value of 0·65 Ω can be checked from Table 8C (ii) of Appendix 8.

**Table 9: Gross cross-sectional areas of armour wires for single-core cables to BS 5467**

| Nominal cross-sectional area of conductor | Cross-sectional area of round armour wires | | | |
|---|---|---|---|---|
| | Cables with stranded conductors | | Cables with solid aluminium conductors | |
| | 600/1000V | 1900/3300V | 600/1000V | 1900/3300V |
| mm² | mm² | mm² | mm² | mm² |
| 50 | 26 | 43 | 23 | 40 |
| 70 | 43 | 48 | 39 | 44 |
| 95 | 48 | 53 | 43 | 48 |
| 120 | 53 | 75 | 47 | 68 |
| 150 | 76 | 81 | 68 | 73 |
| 185 | 84 | 87 | 75 | 78 |
| 240 | 94 | 96 | 84 | 86 |
| 300 | 103 | 105 | 92 | 93 |
| 400 | 145 | 145 | – | – |
| 500 | 161 | 161 | – | – |
| 630 | 180 | 180 | – | – |

In Chapter 6 of this Commentary, a number of graphs are given of the time/current characteristics for various types of overcurrent protective device on which are superimposed the adiabatic lines for a range of copper conductor cross-sectional areas and for $k$ = 115. Although these were developed to illustrate the method of

**Table 10: Gross cross-sectional areas of armour wires for multicore cables to BS 5467 having shaped conductors**

| Nominal cross-sectional area of conductor | Cross-sectional area of round armour wires | | | | | | | | |
|---|---|---|---|---|---|---|---|---|---|
| | Cables with stranded conductors | | | | | Cables with solid aluminium conductors | | | |
| | Two core | Three core | | Four core | | Two core | Three core | | Four core |
| | 600/1000V | 600/1000V | 1900/3300V | Equal neutral 600/1000V | Reduced neutral 600/1000V | 600/1000V | 600/1000V | 1900/3300V | 600/1000V |
| $mm^2$ | $mm^2$ | $mm^2$ | $mm^2$ | $mm^2$ | $mm^2$ | $mm^2$ | $mm^2$ | $mm^2$ | $mm^2$ |
| 16 | 35 | 39 | 73 | 45 | – | 32 | 37 | 69 | 42 |
| 25 | 43 | 63 | 80 | 71 | 71 | 38 | 59 | 76 | 67 |
| 35 | 62 | 70 | 88 | 80 | 77 | 54 | 65 | 82 | 73 |
| 50 | 70 | 79 | 120 | 90 | 88 | 62 | 73 | 113 | 85 |
| 70 | 80 | 91 | 134 | 133 | 129 | 71 | 84 | 125 | 126 |
| 95 | 114 | 130 | 148 | 149 | 147 | 100 | 120 | 138 | 141 |
| 120 | 126 | 143 | 202 | 210 | 165 | – | 132 | 188 | 203 |
| 150 | 139 | 201 | 216 | 232 | 224 | – | 186 | 201 | 220 |
| 185 | 196 | 223 | 234 | 258 | 252 | – | 207 | 216 | 243 |
| 240 | 218 | 250 | 257 | 289 | 280 | – | 230 | 237 | 274 |
| 300 | 239 | 274 | 280 | 319 | 308 | – | 253 | 257 | 300 |

determining compliance with the thermal requirements of *Regulation 434-6* in relation to short circuit protection, they are equally applicable here and, in fact, were the basis for those tables in Appendix 8 of the Wiring Regulations referred to above, for the case where the protective conductor is in the same p.v.c.-insulated cable as the associated phase conductors.

Where it is intended to use the armouring of cables as the protective conductor, the cross-sectional areas necessary for the determination of compliance with *Regulation 543-2* are not readily derived from the values of maximum resistance given in the British Standards for the cables, namely BS 5467 for cables with thermosetting insulation and BS 6346 for armoured p.v.c.-insulated cables.

Tables 8 to 11 give the cross-sectional areas for armouring which have been calculated from first principles and are for round wire armour, generally steel for multicore cables and aluminium for single core cables. The other type of armouring included in the standards for cables with aluminium conductors is aluminium strip, but as its area is more than half the conductor area, Table 54 F of the Wiring

**Table 11: Gross cross-sectional areas of armour wires for single-core cables to BS 6346**

| Nominal cross-sectional area of conductor | Cross-sectional area of round armour wires | | | |
|---|---|---|---|---|
| | Cables with stranded conductors | | Cables with solid aluminium conductors | |
| | 600/100V | 1900/3300V | 600/1000V | 1900/3300V |
| $mm^2$ | $mm^2$ | $mm^2$ | $mm^2$ | $mm^2$ |
| 50 | 40 | 44 | 36 | 41 |
| 70 | 45 | 49 | 41 | 45 |
| 95 | 51 | 72 | 46 | 66 |
| 120 | 73 | 78 | 66 | 71 |
| 150 | 80 | 83 | 73 | 75 |
| 185 | 88 | 90 | 80 | 81 |
| 240 | 99 | 99 | 89 | 89 |
| 300 | 109 | 109 | 98 | 98 |
| 400 | 155 | 155 | – | – |
| 500 | 170 | 170 | – | – |
| 630 | 187 | 187 | – | – |
| 800 | 262 | 262 | – | – |
| 1000 | 289 | 289 | – | – |

Regulations is met and actual calculation for compliance with *Regulation 543-2* is then unnecessary.

**Table 12: Cross-sectional areas of steel conduit**

| Size | Nominal cross-sectional area $mm^2$ | |
|---|---|---|
| | Light gauge | Heavy gauge |
| 16 | 47 | 72 |
| 20 | 59 | 92 |
| 25 | 89 | 131 |
| 32 | 116 | 170 |

**Table 13: Cross-sectional areas of steel trunking**

| Size mm x mm | Nominal cross-sectional area $mm^2$ |
|---|---|
| 50 x 37·5 | 125 |
| 50 x 50 | 150 |
| 75 x 50 | 225 |
| 75 x 75 | 285 |
| 100 x 50 | 260 |
| 100 x 75 | 320 |
| 100 x 100 | 440 |
| 150 x 50 | 380 |
| 150 x 75 | 450 |
| 150 x 100 | 520 |
| 150 x 150 | 750 |

When metallic conduit or metallic cable trunking is used as a circuit protective conductor, the nominal cross-sectional areas based on the standard sizes given in BS 4568 and BS 4678, respectively, are as given in Tables 12 and 13, respectively.

When using the values given in the above Tables, the assumption is made that any joints which are present do not reduce the nominal cross-sectional area. It is also assumed that the lid does not contribute to the cross-sectional area.

One particular type of circuit protective conductor is the combined protective and neutral conductor (called for the sake of brevity the PEN conductor), the requirements for which are prescribed in Section 546 of the Wiring Regulations. The most commonly encountered PEN conductor is of course that in the distribution cable when Protective Multiple Earthing is used and which is usually referred to as a CNE cable, i.e. a combined neutral earth cable. In the installation, the neutral and protective functions are then handled by separate conductors except where earthed concentric wiring is employed.

*Regulation 471-14* indicates that for a circuit incorporating a PEN conductor, although automatic disconnection of the supply can be used as the protective measure against indirect contact, the protective device must not be a residual current device. The reason for this is the simplest: a residual current device, in fact, would not detect earth fault current in these circumstances.

If the PEN conductor is chosen according to *Regulation 546-2* (i), it will be necessary to check that it complies with the thermal requirements of *Regulation 543-2*, and if it is chosen according to *Regulation 546-4*, this effectively is the same as complying with Table 54F and in this case there is no need to carry out any further check.

For protective bonding conductors, whether they are main or supplementary, it is not practicable to calculate the portion of the earth fault current which they may carry. Therefore, Section 547 in prescribing the requirements for these bonding conductors bases their minimum cross-sectional areas on an assessment of probabilities and experience.

The Note to *Regulation 413-2*, which is the basic regulation as regards main equipotential bonding, indicates that compliance with the regulation will normally satisfy the PME Approval but it is important, as stated in the Note to *Regulation 547-2* that, if it is intended to use the PME facility, the supply undertaking concerned should be consulted in order to ascertain if it has any special requirements concerning the size of protective bonding conductors.

It must be remembered that if the consumer cannot comply with the requirements of the PME Approval, then, under the terms of that PME Approval, the supply undertaking is not permitted to connect the installation to a PME earthing terminal and the consumer must then make his own earthing arrangements, usually by installing a residual current device. Such a situation can arise in a household installation where there is a swimming pool. It is then usually impracticable for the consumer to make suitable bonding connections, but if he can: segregate all metalwork associated with the pool from the house by the use of plastics pipes etc.; instal a separate earth electrode connected to the exposed conductive parts of the pool installation; and protect that portion of the installation with a residual current device; then the supply undertaking may allow the use of the PME earthing terminal for the house circuits but only if they were satisfied that the two parts of the installation would remain segregated from each other. The requirements for this segregation would be a minimum of 3 m separation of the pool metalwork from any metalwork associated with the house portion of the installation, including any pipes going into the ground. This can be difficult to confirm and to maintain.

Two further applications where it is unlikely that the consumer can make full use of the PME facility are construction sites and agricultural installations, again because of the impractability of complying with the bonding requirements of the PME Approval, particularly, for the latter, where remote out-buildings are concerned. On construction sites, the supply undertaking may give a PME earthing terminal to site offices if they can be satisfied that it will not be used for earthing purposes on the site.

In agricultural installations the supply undertaking may allow an PME earthing terminal to be used if they are satisfied with the bonding arrangements. In some cases where a transformer supplies only one consumer it is possible to earth its neutral at one point only (at the service point) and provide an earthing terminal. Such a system is known as Protective Neutral Bonding (PNB) and the terms of the PME Approval do not apply, but this system can only be used if there is no possibility of other consumers using that supply.

Chapter 5

# Protection against thermal effects

Chapter 42 of the Wiring Regulations comprises a number of regulations dealing with thermal effects, but, as indicated in the note which precedes that chapter, it does not cover requirements for protection against overcurrent, these being the concern of Chapter 43. Similarly, the chapter does not deal with the thermal aspects associated with earth fault currents (which might also be many times the normal load current) which are covered in Chapter 54. Before commenting on the regulations in Chapter 42, it is felt that it would be instructive to give some indication as to the main sources of fires from electrical causes, at least in domestic dwellings.

In 1976, when electrical safety in the home was the responsibility of the former Department of Prices and Consumer Protection (that responsibility now rests with the Department of Trade) a 'Commentary on electrical fatalities in the home 1974-75' was published, which included a table giving the number of fires from electrical causes in dwellings for the years 1965 to 1972. That table is reproduced here as Table 14 and it will be seen that fires from electrical causes increased from approximately 12 000 to 21 700 in that period.

The DPCP Commentary pointed out that the fires attributed to defective wiring included those due to flexible cords, and that many houses had insufficient points for the vastly increased number of appliances in use. It also indicated that the steady figures for wiring were encouraging as they showed what must be an improving situation. Most of the fires attributable to electric cookers were chip pan fires and the main culprit in the item 'space heating' was the radiant portable fire.

Particular reference was made to the number of 'airing cupboard' fires involving immersion heaters, and for the three years 1970/71/72 the figures clearly showed an unsatisfactory situation. These fires were believed to have resulted from the continuous operation of cables and the use of the 13A plug and socket outlet in a high ambient temperature, the view being expressed that this was not good practice. In Chapter 8 of this present Commentary, there is some consideration given to this

Table 14: **Fires from electrical causes in dwellings**

| | *1972* | *1971* | *1970* | *1969* | *1968* | *1967* | *1966* | *1965* |
|---|---|---|---|---|---|---|---|---|
| All fires in dwellings | 52 868 | 45 955 | 45 305 | 45 872 | 43 072 | 39 266 | 34 931 | 34 549 |
| Cookers | 11 296 | 9 512 | 8 311 | 7 696 | 6 816 | 5 889 | 5 063 | 4 424 |
| Space heating | 2 021 | 1 551 | 1 806 | 1 958 | 1 560 | 1 502 | 1 515 | 1 501 |
| Central heating | 47 | 33 | 43 | | | | | |
| Water heating | 950 | 764 | 644 | | | | | |
| Wiring | 3 203 | 2 951 | 3 191 | 3 456 | 3 000 | 3 083 | 2 904 | 2 928 |
| Welding | 4 | 3 | 5 | | | | | |
| Lighting | 238 | 228 | 278 | 242 | 200 | 170 | 185 | 162 |
| Blanket | 1 499 | 1 333 | 1 432 | 1 624 | 1 552 | 1 511 | 1 459 | 1 365 |
| Radio and TV | 1 946 | 1 910 | 1 716 | 1 436 | 1 328 | 1 227 | 1 100 | 945 |
| Refrigerators | 86 | 54 | 42 | 48 | 60 | 61 | 39 | 68 |
| Supply apparatus | 28 | 25 | 28 | 34 | 24 | 24 | 49 | 66 |
| Other | 416 | 321 | 370 | 986 | 872 | 820 | 690 | 624 |
| Total | 21 733 | 18 685 | 18 066 | 17 480 | 15 412 | 14 287 | 13 004 | 12 087 |
| Per cent of all fires | 41 | 45 | 40 | 38 | 35 | 36 | 37 | 35 |
| Excl Cookers | 10 437 | 9 173 | 9 555 | 9 784 | 8 596 | 8 398 | 7 941 | 7 653 |
| Per cent | 20 | 20 | 21 | 21 | 20 | 21 | 23 | 22 |

By courtesy of the Department of Trade

particular aspect where it is emphasised it is essential that all equipment, not only plugs and socket outlets, must be suitable for the environmental conditions likely to be encountered.

It is clear from the evidence of Table 14 that a considerable percentage of the fires attributed to electrical causes are, in fact, attributable to portable equipment and neither the installation designer nor contractor has any control over the selection of such equipment or the manner in which it is used.

All the regulations in Chapter 42, for that very reason, are concerned only with *fixed* equipment and *Regulation 422-1* is generally applicable to such equipment. British Standards dealing with the safety of current-using equipment such as BS 3456 for household appliances, BS 3861 for office machines and BS 5784 for commercial catering equipment, include a heating test in which the equipment is installed in a standardized test corner in a specified position or in accordance with instructions by the manufacturer concerning its installation. The temperature rises of all components, insulation and motor windings are checked as are those of the wooden supports, walls, ceiling and floor of the test corner itself and these must not exceed specified maximum values. With regard to the temperature rises associated with the test corner, these maximum temperature rises are 60°C if the equipment being tested is intended to be operated continuously for long periods or 65°C for other equipment, the reference ambient temperature being 20± 5°C.

These British Standards include an abnormal operation test, also carried out in the test corner during which the temperature rises of the parts of that corner shall not exceed 150°C and this test is intended to check that the design of the equipment is such that the risk of fire (and of mechanical damage and the loss of protection against electric shock) as the result of abnormal or careless operation is obviated as far as possible.

When an installation includes fixed equipment, the fact that the selection of the equipment may have been made by other than the installation designer or contractor nevertheless does not absolve him from the responsibility of seeing that any installation instructions of the manufacturer are adhered to, and that any structural materials adjacent to the installed equipment do not represent a fire hazard, bearing in mind the possible temperatures which will be attained by that equipment.

*Regulation 422-4* which concerns fixed luminaires and lamps should be read in conjunction with *Regulation 553-16*. In essentials, Appendix A to BS 5042 Part 1, to which reference is made in the Note to *Regulation 553-16*, suggests that for other than 60W lamps in B15 lampholders and 200W lamps in B22 lampholders (except in the cap-up position with an open-type shade) the temperature rating of the lampholder may be TI for single-lamp luminaires or for multilamp luminaires having widely spaced lamps. Where lamps are in close proximity to each other or are in totally enclosed luminaires T2 lampholders should be used. The reader is reminded that for the purposes of the Wiring Regulations a batten lampholder and a lampholder suspended by a flexible cord are considered to be luminaires. Table 4A of Appendix 4 of the Wiring Regulations allows the designer to assume, in the absence of contrary information, that a lighting outlet will supply a minimum of 100 W per lampholder. This would appear to allow batten lampholders and lampholders suspended by a flexible cord to be of TI temperature rating.

A great deal of work is being undertaken both nationally and internationally on the subject of protection against fire and Section 422 of the Wiring Regulations will undoubtedly be enlarged when agreement has been reached on the various aspects now under discussion.

The Wiring Regulations contain a number of individual regulations, apart from those in Section 422, which are intended to provide protection against thermal effects. Section 528 covers the provision of fire barriers and *Regulation 523-6* requires internal barriers in vertical channels, ducts, ducting and trunking containing conductors or cables. The whole of Chapter 43 of the Wiring Regulations which prescribes requirements for overcurrent protection is based on seeing that under overload and short circuit conditions the temperatures attained by conductors and their insulation are limited and that the protective devices are themselves capable of withstanding the thermal stresses created by overcurrents. The comparative brevity of Section 422 may therefore be misleading as to the efforts that are made in the Wiring Regulations to provide safety of persons and of property from fire.

The user of the Wiring Regulations will notice that some of the fire technology terms used in previous editions are no longer present, in particular, 'non-combustible'. The reason for this change in terminology is that in the mid-1970s the British Standards Institution set up a Coordinating Committee on Fire Tests and one of the first tasks of that Committee was to examine the situation that existed at that time in respect of the terminology used in fire tests prescribed in numerous British Standards, including those concerning electrical products.

An analysis of those fire tests showed anomalies and the inappropriate use of words describing fire tests, the results of those fire tests and the fire properties of the materials concerned. It was felt by the Coordinating Committee that there could be the possibility of confusion and misconception in the technology being used which in turn could lead to danger.

In the preparation of the Fifteenth Edition it was decided that due account should be taken of the guidelines prepared by the BSI Coordinating Committee on Fire Tests. All the fire terms used in the Wiring Regulations were considered and, wherever they appeared, any deprecated terms such as 'non-combustible' were removed, being replaced by the appropriate acceptable terms.

Many British Standards covering the safety requirements for electrical products include tests intended to determine the resistance to heat, fire and tracking of those products. The details of these tests need not be considered here, but generally they are the so-called ball-pressure test, glow-wire test, a tracking test, and others.

Conduit, trunking and cable-ducting product standards include similar tests for fire hazard assessment, and these are under review both nationally and internationally to ensure a proper correlation between these and other electrical products associated with them and the building materials with which they are in close contact after installation. Basically there are two requirements for cables in enclosures of the types mentioned; first, they must not constitute a source of ignition hazard and, secondly, they must not propagate a fire which has been caused elsewhere but which may engulf them as it develops.

So far as the ignition hazard is concerned, a correctly designed wiring system, i.e. properly selected cables installed in equally properly selected conduit or other enclosure, should not cause ignition hazards except where mechanical damage has penetrated both the enclosure and the cable insulation and where, in addition, the protective device(s) associated with the circuit(s) concerned has failed to operate.

The other possible cause of ignition hazard occurs if a cable joint within the wiring system develops a fault leading to overheating and this, of course, is why joints, with few exceptions, are required to be accessible for inspection. The accessibility of joints is also necessary for ease of testing.

As regards the fire propagation hazard, this results from a fire initiated elsewhere in the building (i.e. not in the electrical installation). In this case, the cables and their enclosures are a source of fuel, adding either substantially or negligibly to the total fuel available to sustain the fire. It may be that the cables and their enclosures are required to penetrate fire barriers and for this reason, in order not to reduce the integrity of such barriers, *Regulation 528-1* requires properly constructed fire stops around the cable enclosures and the provision of suitable internal fire barriers. Currently, there are tests prescribed for individual and bunched cables and for enclosure systems in relation to the propagation of flame and certain tests are being devised to interpret the bunched cable performance when bunches of cables are enclosed in conduit or trunking.

In general terms, cables and their enclosures complying with these tests are regarded as not contributing to the propagation of flame but nevertheless, in a well developed fire, the wiring system may be totally consumed together with the surrounding building materials, flooring and furnishings. In such a situation, the contribution by the wiring system is generally insignificant except in very specialised buildings such as telephone exchanges or power stations where, in any event, extra requirements are already specified.

Chapter 6

# Protection against overcurrent

Chapter 43 of the Wiring Regulations prescribes the fundamental requirements for the protection of circuits against overcurrent but it has to be read in conjunction with Section 473, the regulations in that section indicating the application of the protective devices giving overcurrent protection. Section 473 also indicates the circumstances in which it is permissible to dispense with overcurrent protection while Section 533 prescribes requirements for the overcurrent protective devices used.

The term 'overcurrent' includes both overload and short circuit currents where overload current is defined as an overcurrent occurring in a circuit which is perfectly sound and is not to be regarded as due to an electrical fault, e.g. an overcurrent due to a stalled motor. Short circuit current is that which results from a fault of negligible impedance between live conductors normally having a difference in potential but the Wiring Regulations deal only with short circuits occurring between live conductors of the same circuit. (In the Fifteenth Edition, the neutral conductor of an a.c. circuit or its equivalent in a d.c. circuit is considered to be a live conductor).

Note 2 to *Regulation 431-1* warns that compliance with the requirements concerning overcurrent protection for a particular circuit does not mean that such protection is then automatically extended to the current-using equipment fed by that circuit or to the flexible cables or cords connecting the equipment to the circuit through plugs and socket outlets. The latter aspect is of particular importance when non-fused plugs are used, such as BS 4343 industrial type plugs, and if the designer intends to use a circuit incorporating these, or a similar plug and socket outlet system (other than the standard circuit detailed in item C in Appendix 5 of the Wiring Regulations), he should establish, if possible, the sizes and types of flexible cables and cords likely to be used and assure himself that these will be protected by the overcurrent protective device he intends to use for the circuit. It is most unlikely that the designer will be able to obtain this information and in such cases it is necessary to limit the current rating of the circuit to such a value that the overcurrent protective device *does* give protection to even the smaller sizes of flexible cords.

## 6.1 Basic requirements for protection against overload

The basis used for the requirements for protection against overload is the limiting temperature of the insulation of the conductors of the installation when subjected to overloads of fairly long duration, of the order of 1 hour or more, and these requirements, which are prescribed in *Regulation 433-2*, are intended to see that the life of the conductor insulation is not significantly shortened. In Chapter 9 of this Commentary, in the comments on *Regulation 522-1*, the maximum permissible operating temperatures of various insulating materials are given, together with comments on cable current-carrying capacities and how these are affected by ambient temperature and cable grouping.

*Regulation 522-1* requires that the current-carrying capacity $I_Z$ of a cable conductor is not to be less than the maximum sustained current normally carried by the conductor, this current generally being taken to be the design current $I_B$ of the circuit. Thus $I_B \leqslant I_Z$ which is also the net result of combining items (i) and (ii) of *Regulation 433-2*, that combination being shown symbolically in Note 1 to the regulation as $I_B \leqslant I_n \leqslant I_Z$ where $I_n$ is the nominal current rating or current setting of the overload protective device.

Fuses, of course, do not 'blow' at their rated currents and neither do miniature circuit breakers (or similar devices) trip at their rated currents. For instance, a general purpose fuse to BS 88 Part 2 having a rated current equal to or less than 63A is required *not* to operate when it carries a current 1·2 times its rated current for 1 h. This time of 1 h is termed the 'conventional time' and for higher rated currents increases to 4 h. This type of fuse is required to operate within its conventional time when it carries a current 1·6 times its rated value.

Similarly, miniature circuit breakers to BS 3871 are required by the British Standard not to operate when continuously carrying their rated currents, and while a maximum value of non-tripping current is not specified as such they are required to trip within 1 h if the current is 1·5 times rated current (for rated currents up to and including 10 A) or 1·35 times rated current (for rated currents greater than 10A.) There are international proposals which, if agreed, could be included in a revision of BS 3871 and these would require that a miniature circuit breaker having a rated current not exceeding 63 A shall not trip within 1 h (conventional time) when carrying a current of 1·13 times rated current but must trip within the following hour (again this is the conventional time) if the current is then increased to 1·45 times rated current. For rated currents exceeding 63 A it is proposed that the conventional time becomes 2 h. Equivalent figures for circuit breakers of the conventional or moulded case type are given in Table VIII of the relevant standard (BS 4752 Part 1).

The designer must *not* take advantage of the fact that the overload protective device may be able to carry a current somewhat greater than its rated value for a significant period (or even continuously) and as indicated in Note 2 to *Regulation 433-2* the Wiring Regulations do not envisage the occurrence of frequent overloads of long duration but which are sufficiently small so that the device does not operate. Although such overloads may not be directly dangerous, they can shorten the life of the insulation of the conductors concerned and it is important that the design current $I_B$ of a circuit is accurately calculated and if diversity is applied it is done with care. If small overloads are thought likely to be frequent they should be treated not as overloads but as the normal current of the circuit.

The Wiring Regulations, in item (iii) of *Regulation 433-2*, recognise the need to protect circuits from heavier overloads and the current assuring effective operation of the protective device $I_2$, referred to in that item, is, in fact, taken to be the current causing operation within conventional time with some allowance made for the test methods used to determine that current. Thus although the latter is 1·6 times rated current for a BS 88 fuse and 1·5 times rated current for a BS 3871 miniature circuit breaker as compared with the 1·45 times quoted in item (iii), it is accepted that such devices meet the requirement of item (iii) without further proof provided the circuit design meets the requirements of item (ii).

However, semi-enclosed fuses to BS 3036 may have a fusing factor of 2, fusing factor being defined as the ratio of minimum fusing current for operation within 4 hours to rated current.

Thus, for such fuses, compliance with item (ii) of *Regulation 433-2* no longer gives automatic compliance with item (iii) and it becomes necessary to proceed as follows.

As

$$I_2 = 2I_n$$

then, to meet item (iii)

$$2I_n \leqslant 1{\cdot}45\, I_Z$$

from which

$$I_n \leqslant 0{\cdot}725\, I_Z$$

This means that when it is intended to protect a circuit by means of semi-enclosed fuses a larger cable than that determined from normal load conditions must be selected, for any given nominal rated current of fuse, and this need to increase the cross-sectional area of cable conductors applies irrespective of the type of cable being protected (except mineral-insulated cables).

It is therefore seen that the distinction between 'close' and 'coarse' overcurrent protection used in the Fourteenth Edition is no longer made in the same form. The current-carrying capacity of any cable or conductor *under normal load conditions* is affected by the ambient temperature in which the cable or conductor is operating,

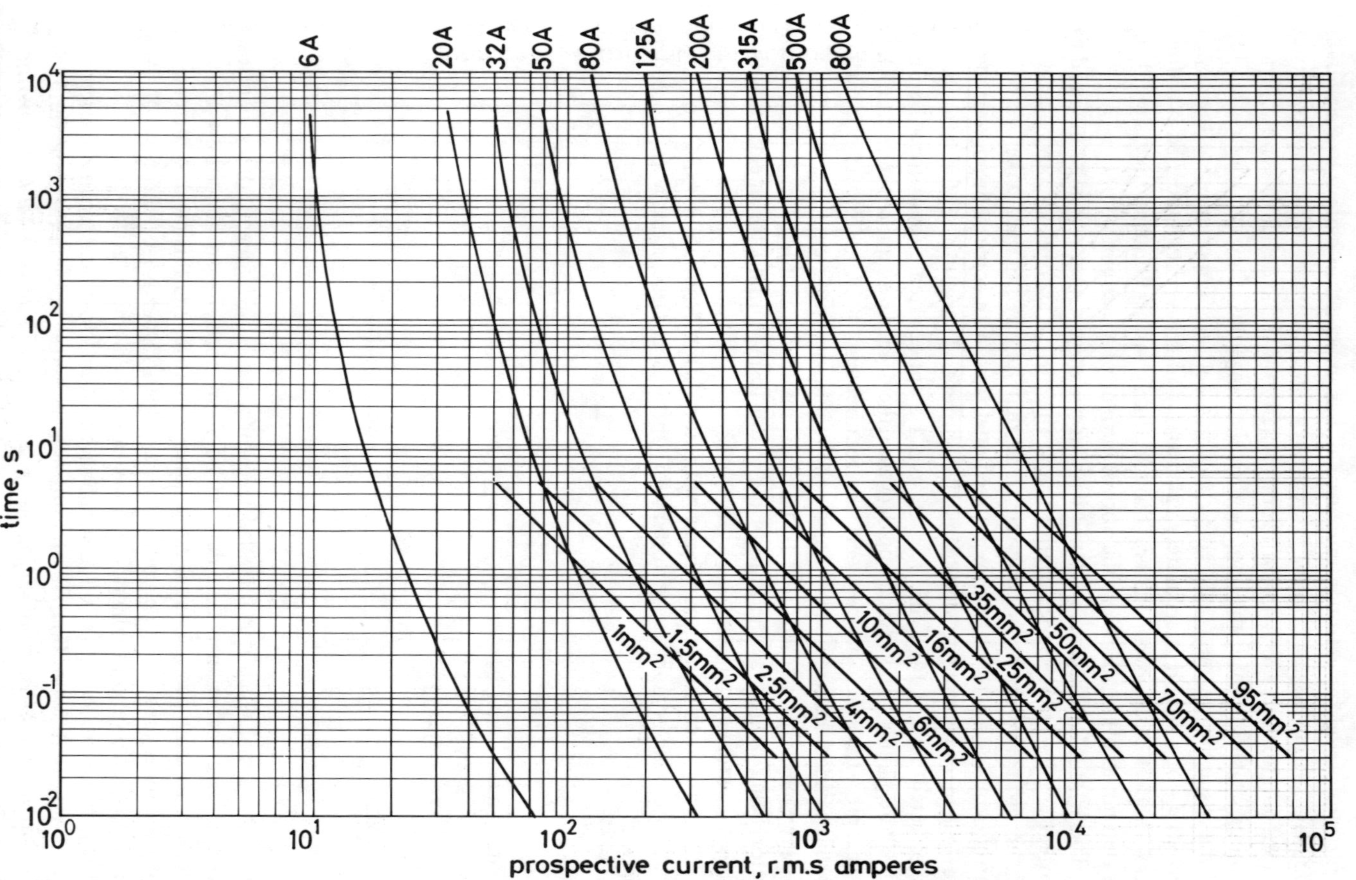


**Fig. 13** *Time/current characteristics for fuses to BS 88 Pt. 2 and adiabatic lines for copper conductors when k = 115*

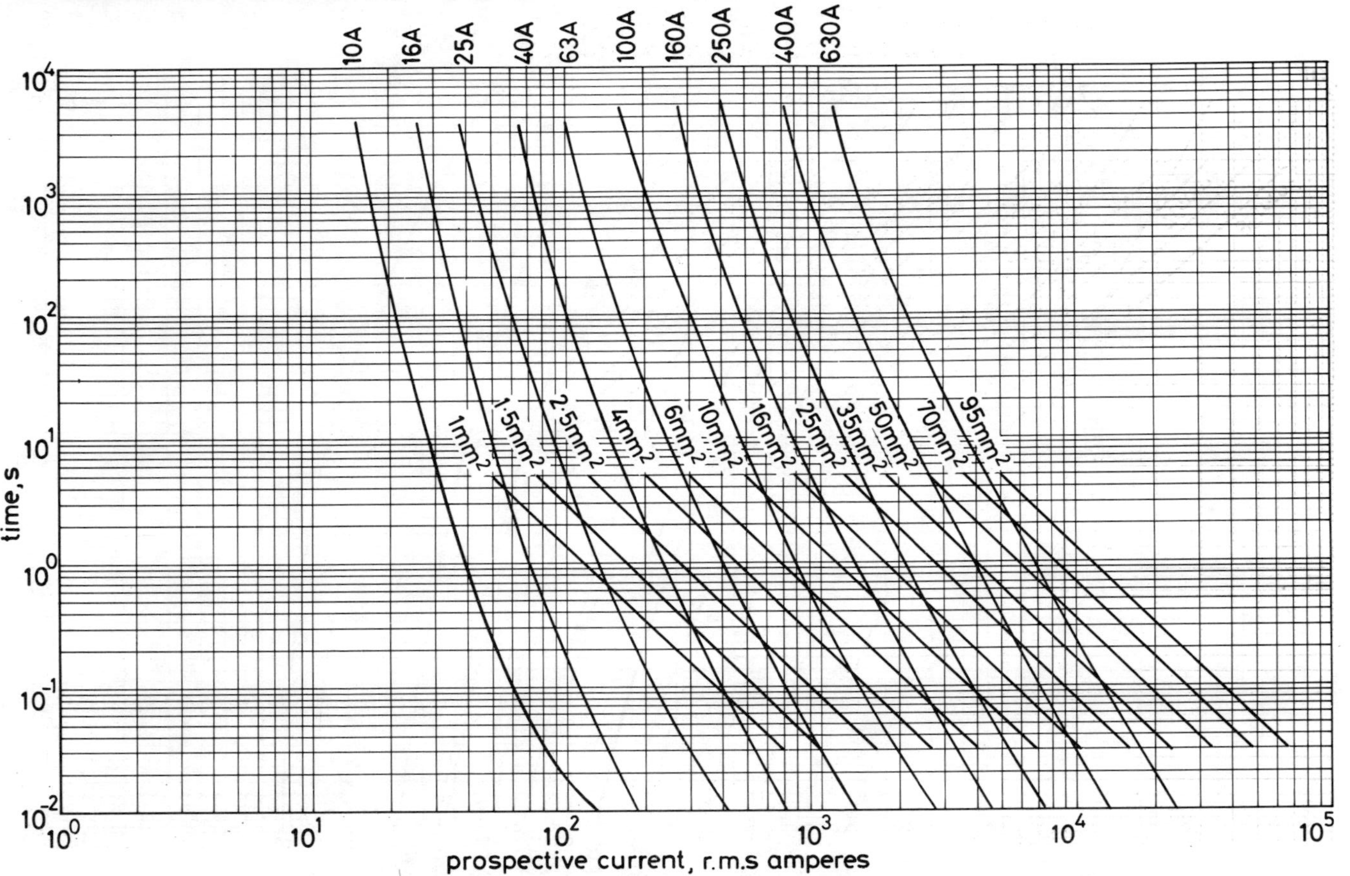


**Fig. 14** *Time/current characteristics for fuses to BS 88 Pt. 2 and adiabatic lines for copper conductors when k = 115*

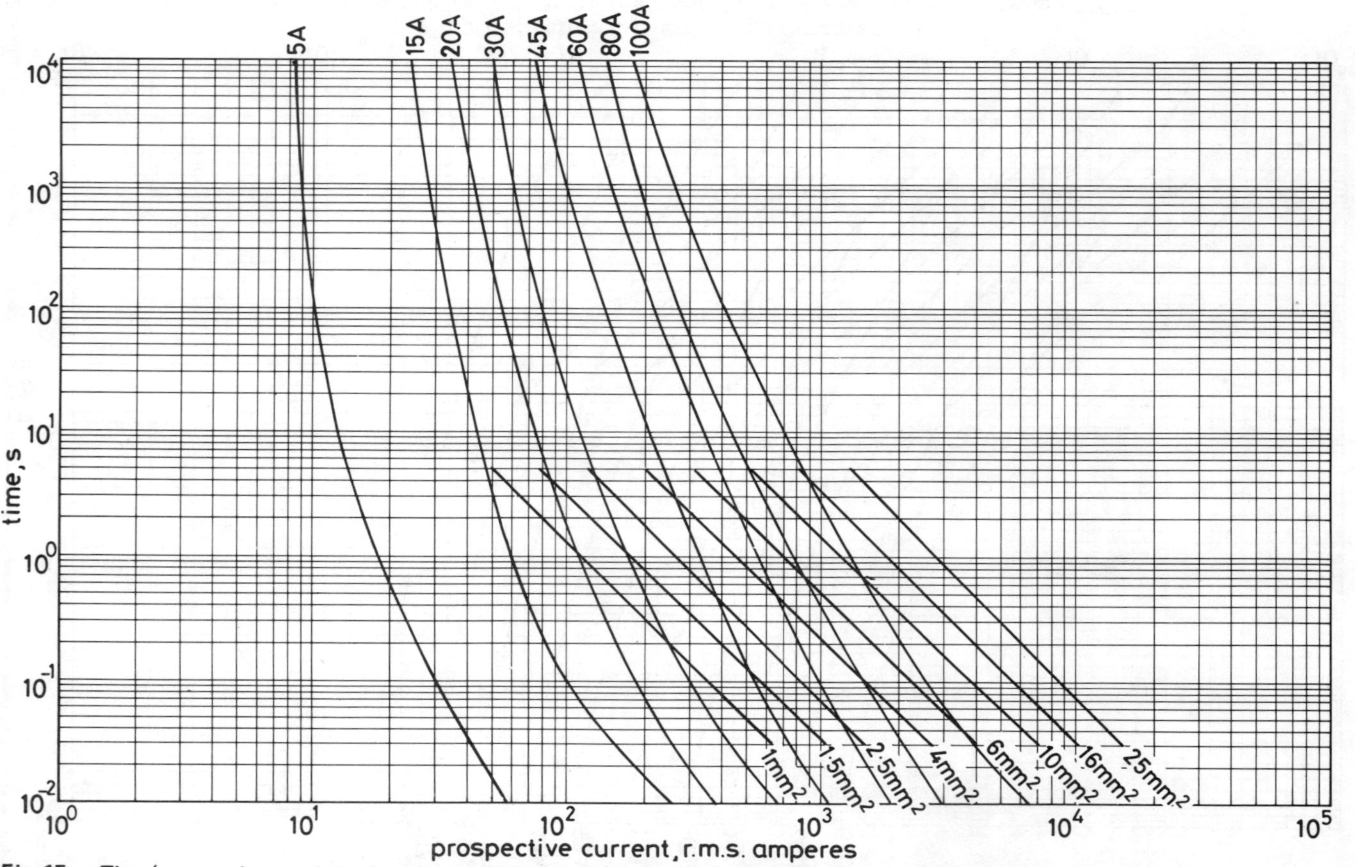


**Fig. 15** *Time/current characteristics for fuses to BS 1361 and adiabatic lines for copper conductors when k = 115*

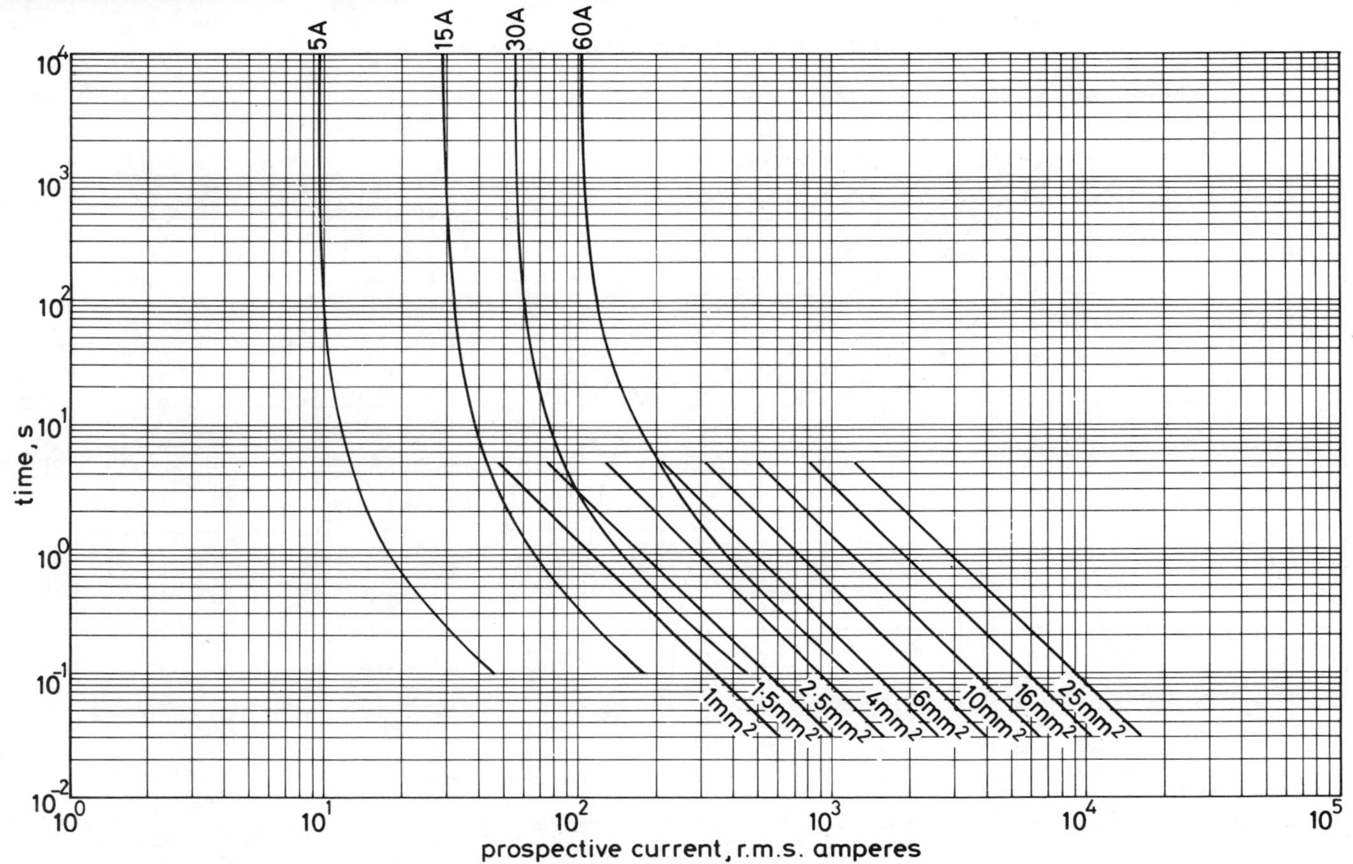


**Fig. 16** *Time/current characteristics for semi-enclosed fuses to BS 3036 and adiabatic lines for copper conductors when k = 115*

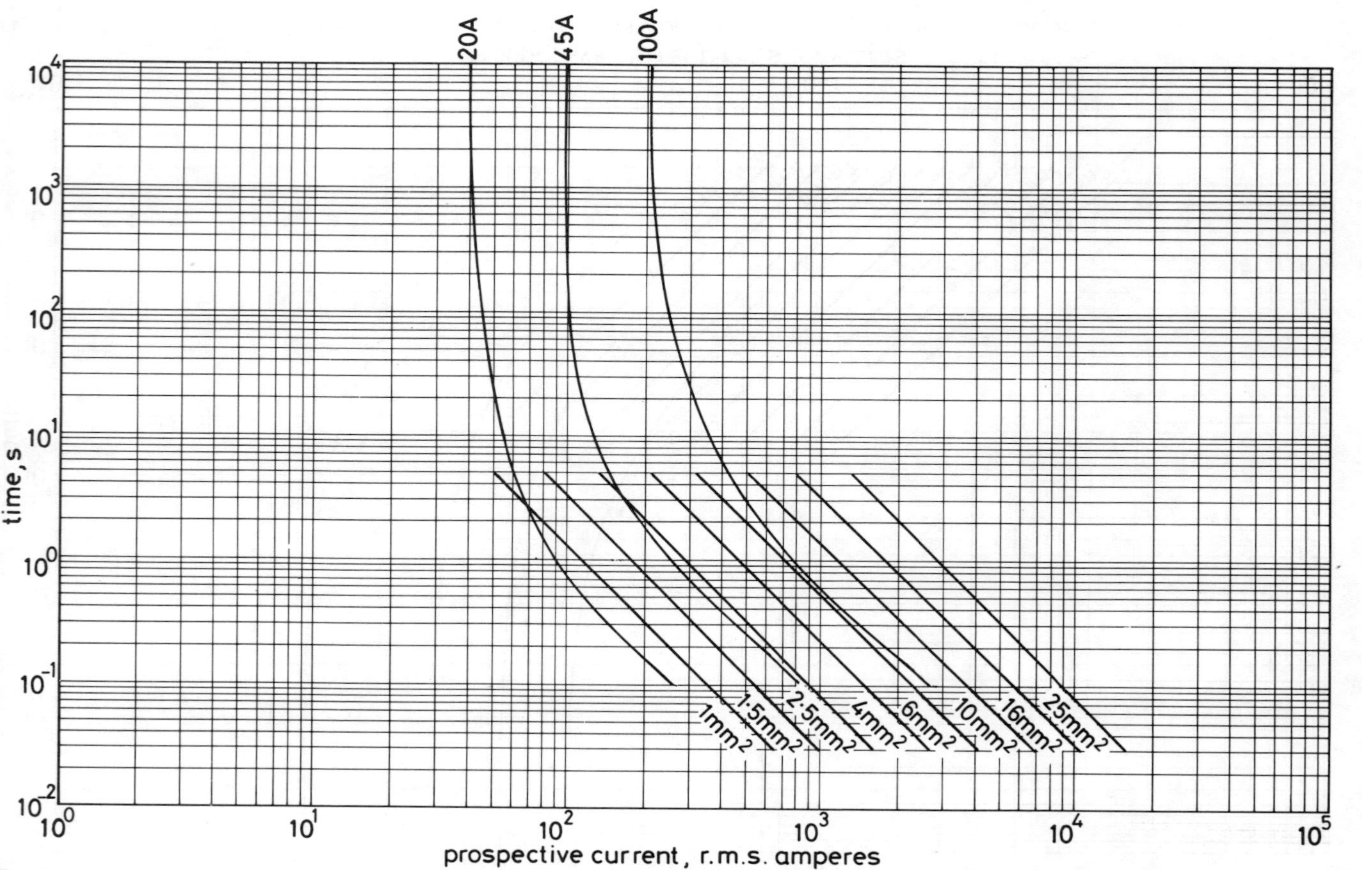


**Fig. 17** *Time/current characteristics for semi-enclosed fuses to BS 3036 and adiabatic lines for copper conductors when k = 115*

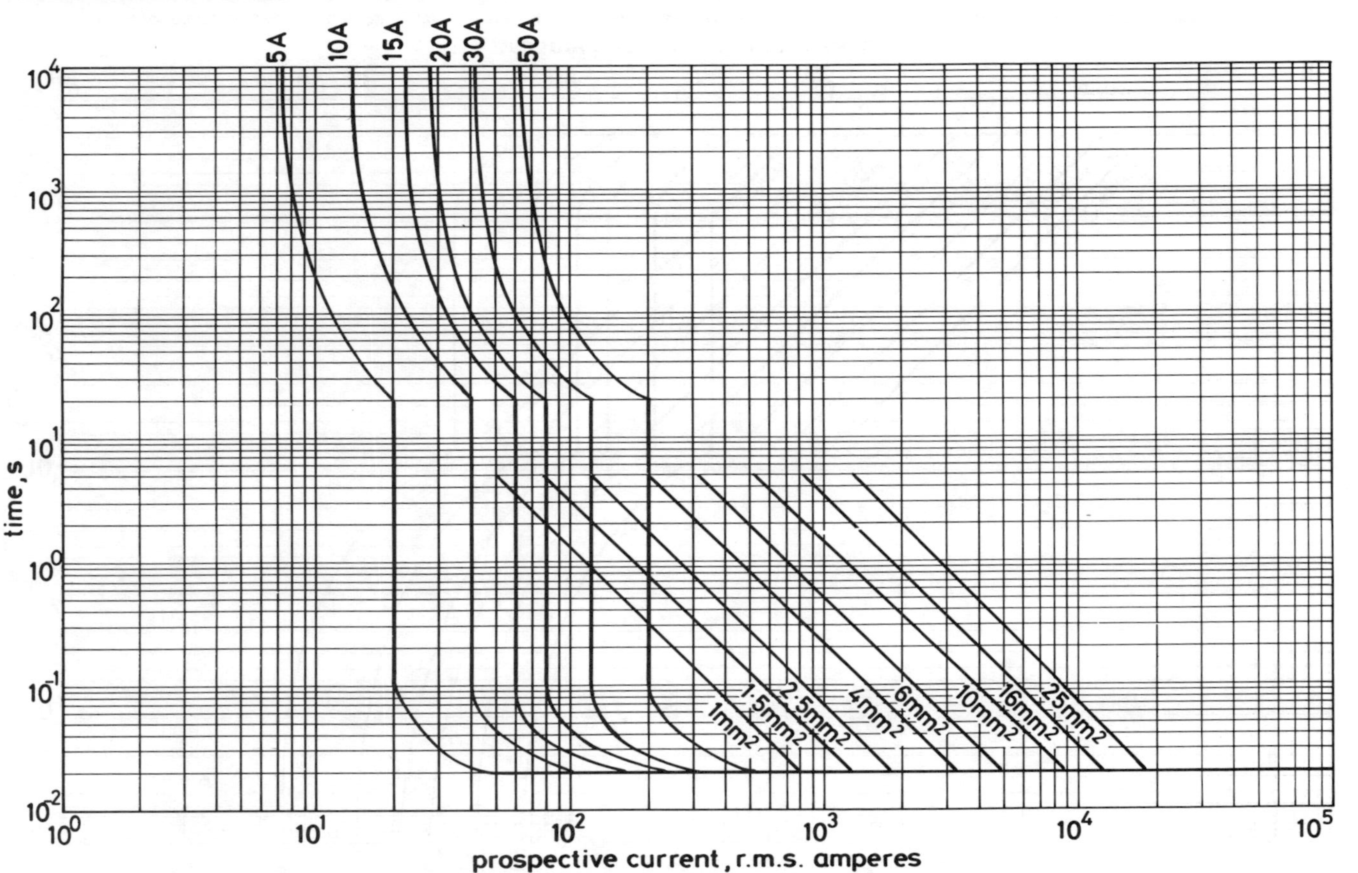


**Fig. 18** *Time/current characteristics for Type 1 miniature circuit breakers and adiabatic lines for copper conductors when k = 115*

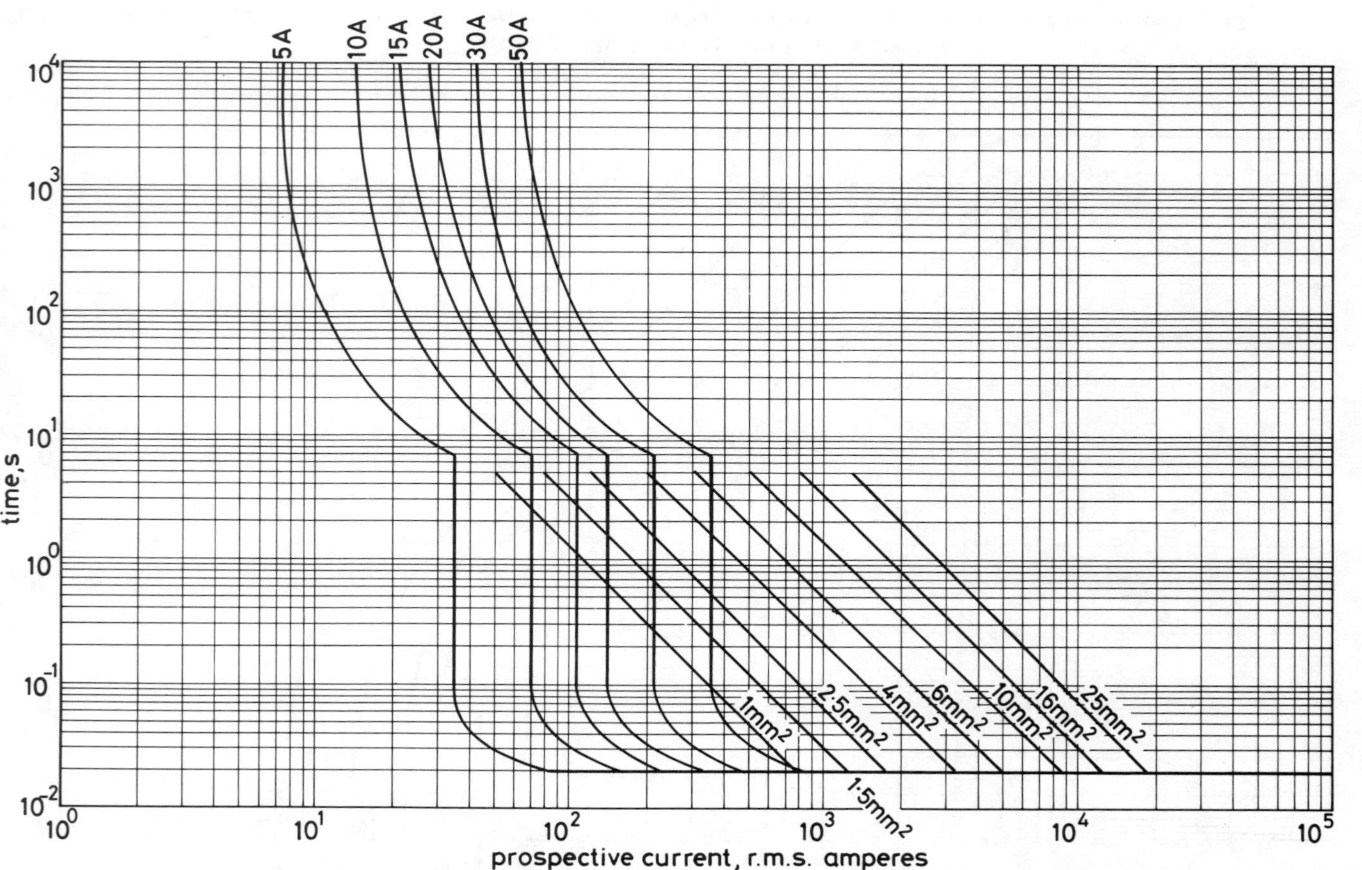


**Fig. 19** *Time/current characteristics for Type 2 miniature circuit breakers and adiabatic lines for copper conductors when k = 115*

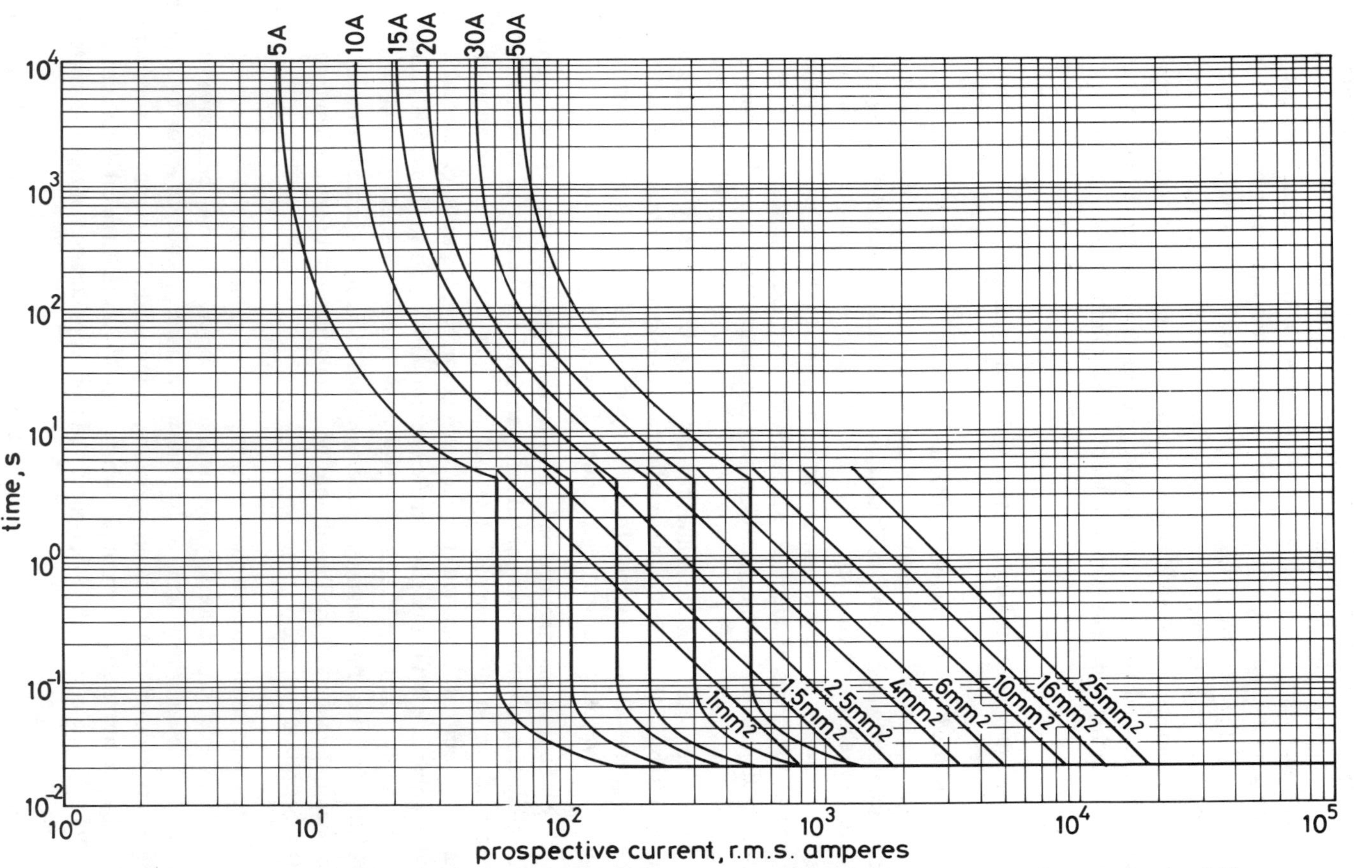


**Fig. 20** *Time/current characteristics for Type 3 miniature circuit breakers and adiabatic lines for copper conductors when k = 115*

the method of installation, and on whether it is grouped or bunched with cables of other circuits but it is *not* affected by the type of overcurrent protection it is intended to use. The 'derating' of a circuit because it is to be protected by a semi-enclosed fuse is due solely to consideration of the overload currents (both magnitude and duration), and hence of the cable temperatures which could occur because of the relatively high fusing factor and conventional time of the semi-enclosed fuse. It was therefore considered that the most logical approach was to give in Appendix 9 of the Wiring Regulations tables of current-carrying capacities which correspond to continuous loading, i.e. of full thermal current ratings, but these are further commented on in relation to *Regulation 522-1*.

## 6.2 Basic requirements for protection against short circuit

The basis used for the requirements for protection against short circuit is the equation, the so-called adiabatic equation, given in *Regulation 434-6*. This is the same equation which appears in *Regulation 543-2* but expressed differently because, whereas in relation to *Regulation 543-2* it is used to determine the minimum cross-sectional area of protective conductors or, if using a standard cable, that the protective conductor is of adequate cross-sectional area, here the equation is generally used to determine whether the disconnection time attained by the short circuit protective device is sufficiently rapid to prevent excessive conductor temperatures. It should be noted that the values for the factor $k$ are exactly the same as those given in Table 54C for protective conductors as a core of a cable and this table tells the designer both the assumed initial temperature (at the commencement of the fault) and the permitted final temperature of the conductor.

It will be seen that the assumed initial temperature is the same as the maximum operating temperature under normal load permitted for the type of cable insulation concerned, or in other words, it is assumed that the cable is operating at the maximum current permitted by the current-carrying capacity of the cable. It follows that the $k$ factors given in *Regulation 434-6* (and in Table 54C) do not strictly apply when the overcurrent protective device is a semi-enclosed fuse to BS 3036 because, for instance, a p.v.c.-insulated cable in these circumstances will be operating under full load conditions with a conductor temperature less than the 70°C permitted for p.v.c.

It was felt unnecessary to add a further set of values for the factor $k$ as this would have been an unnecessary elaboration. However, there may be circumstances where the designer may wish to be able to calculate a value of $k$ for an initial temperature other than that corresponding to the maximum permitted conductor operating temperature, for example, where it is known that the cable concerned is carrying far less current than that permitted from consideration of that conductor's current-carrying capacity.

Without elaborating on the theme, the factor $k$ is derived from the following formula, taken from the IEC Standard corresponding to Chapter 54 of the Wiring Regulations.

$$k = \sqrt{\frac{Q_c\,(B + 20)}{Q_{20}} \ln \left(1 + \frac{\theta_f - \theta_i}{B + \theta_1}\right)}$$

where

$Q_c$ = volumetric heat capacity of conductor material (J/°C mm³)
$B$ = reciprocal of temperature coefficient of resistivity at 0°C for the conductor (°C)
$Q_{20}$ = electrical resistivity of conductor material at 20°C (Ω mm)
$\theta_1$ = initial temperature of conductor (°C)
$\theta_f$ = final temperature of conductor (°C)
ln = $\log_e$

| Material | $B$ °C | $Q_c$ J/°C mm³ | $Q_{20}$ Ω mm | $\sqrt{\frac{Q_c(B+20)}{Q_{20}}}$ |
|---|---|---|---|---|
| Copper | 234.5 | $3.45 \times 10^{-3}$ | $17.241 \times 10^{-6}$ | 226 |
| Aluminium | 228 | $2.5 \times 10^{-3}$ | $28.264 \times 10^{-6}$ | 148 |
| Lead | 230 | $1.45 \times 10^{-3}$ | $214 \times 10^{-6}$ | 42 |
| Steel | 202 | $3.8 \times 10^{-3}$ | $138 \times 10^{-6}$ | 78 |

The magnitude of the prospective short circuit current at the origin of an installation fed by the public supply network is limited only by the impedances of the phase and neutral conductors of the supply cables and of the source of energy. It follows that the prospective short circuit current can only be determined by reference to the supply undertaking and as already indicated in Chapter 3 of the Commentary, the supply undertaking will usually only be able to give a range of probable values.

Fig. 10 in Chapter 4 of this Commentary showed how one could determine the minimum earth fault current to satisfy the thermal requirements of *Regulation 543-2* and clearly the same approach can be used when considering short circuit currents. Design curves can be readily produced by superimposing on the time/current characteristic of any overcurrent device a family of adiabatic lines for various cross-sectional areas but for a particular value of $k$.

This method has been adopted for the following series of graphs (Figs. 13 to 20) based on the time/current characteristics given in Appendix 8 of the Wiring Regulations and for $k = 115$, this value of $k$ being for p.v.c.-insulated copper conductors. If these graphs are used to determine the cross-sectional areas of protective conductors the value of $k$ chosen is for the case where a protective conductor is part of the same cable as the associated phase and neutral conductors.

It must be noted that the time/current characteristics in Appendix 8 and repeated here are the maximum permitted disconnection times; actual characteristics may be faster.

Whereas the characteristic curves for fuses are clearly defined by the relevant British Standards (except for the semi-enclosed fuse to BS 3036) the standard for miniature circuit breakers recognises four different types of overload characteristic, depending on the point where the curve ceases to have an inverse time characteristic and apparently becomes instantaneous. (Tripping in a time less than 0·10 s is considered instantaneous).

Type 1 offers instantaneous tripping between 2·7 and 4 times full load current and such breakers are particularly suited to give shock risk protection against indirect contact in addition to overload protection. However, they may be unsuitable for use on circuits with appreciable inrush currents and type 2 (4 to 7 times full load current) or Type 3 (7 to 10 times full load current) would be preferable for circuits feeding banks of lighting, motors, transformers etc. Type 4 (in excess of 10 times full load current) may be necessary for X-ray machines, welding equipment etc. All types of miniature circuit breakers will give adequate short circuit protection to cables, but the degree of protection can be improved by using the type with the lowest instantaneous trip current compatable with the inrush currents likely to develop.

## 6.3 Protection against overload

Having commented on the basic requirements for both overload and short circuit protection it now becomes necessary to consider the other regulations in Chapter 43 and those in Section 473.

As regards overload protection, for a simple circuit supplying only one item of equipment, it makes no difference electrically whether the overload protective device is fitted at the point of origin of the circuit, or along its run, and this is recognised in *Regulations 473-1 and 473-2*. It is quite common practice for the overload device to be placed towards the load end of the circuit, for example, in a motor starter incorporating an overload release. As indicated in *Regulation 473-2* the part of the circuit between the origin and the overload protective device must not incorporate any branch circuits or socket outlets and must be protected against short circuit.

*Regulation 432-3* allows the overload protective device to have a breaking capacity below the value of the prospective short circuit current at the point where it is installed and the example just quoted is a typical case. In these motor circuits, back-up protection, i.e. the short circuit protection, is provided by, for instance, HBC fuses or miniature circuit breakers and the cross-sectional area of the conductors between the short circuit protective device and the overload protective device is determined by the full load current (where necessary, also taking account of starting currents) and the requirements for limitation of voltage drop. The current rating of the short circuit protective device, however, can be selected to suit the starting characteristics and a rating of about twice the full load current is usually suitable unless the starting conditions are severe. This is recognised by *Regulation 434-1* but it is still necessary to establish that the disconnection time of that short circuit protective device will operate sufficiently rapidly in the event of a fault in order that the temperature of the conductors does not exceed the permitted maximum.

*Regulation 473-3* indicates the conditions under which it is admissible to dispense completely with overload protection and, as regards item (ii) of that regulation, fixed heating appliances are examples of equipment unlikely to cause overload current. Although such appliances may not be likely to cause overload they may be subject to fault conditions which would cause overcurrent of a similar magnitude to overload, a point recognised in Note 1 to *Regulation 431-1*, and this risk must be borne in mind where it is intended to omit overload protection. Circuits incorporating socket outlets must always be regarded as susceptible to overload and therefore *Regulation 473-3* cannot be applied to such circuits.

## 6.4 Protection against short circuit

The subject of short circuit protection is certainly more complex than that of overload protection, and in this Commentary it is only possible to give a very brief outline of it. Mention has already been made of the breaking capacity of a protective device and of 'prospective short circuit current' and Section 434 also refers to 'asymmetry', 'let-through energy' and 'current-limiting'. It is felt that, at least, a brief explanation of these terms is appropriate and that explanation, it is believed, will go some way to also explaining the intent of the regulations in Section 434.

The prospective short circuit current at a point in a circuit can be defined as the r.m.s. symmetrical current which would flow in that circuit if a link of negligible impedance connected the phase conductor(s) and neutral conductors together at that point, i.e. the link created a 'dead short'.

The rated breaking capacity of a fuse is expressed in amperes and is defined as the maximum prospective current that the fuse is capable of breaking under specified conditions. If one is using a miniature circuit breaker or similar, the corresponding term is 'short circuit capacity' because the device must be able to both make and break the prospective short circuit current.

The key requirement as regards short circuit protection is that given in *Regulation 434-4*, namely that the breaking capacity rating of a protective device intended to give such protection must not be less than the prospective short circuit current at the point where it is intended to instal the device and Table 15 gives the breaking capacity ratings of commonly used types of protective device.

The Note to *Regulation 13-7* indicates that for installations where the supply undertaking provides switchgear at the origin of the installation there may be no

**Table 15: Short circuit breaking capacities of types of protective device**

| Type of device | Breaking capacity rating kA | At nominal power factor |
|---|---|---|
| Fuses to BS 1361, Type I | 16.5 | 0.3 |
| Fuses to BS 1361, Type II | 33.0 | 0.3 |
| Fuses to BS 88 | To be ascertained from the manufacturer's data, for the type reference number marked on the fuse link. Breaking capacity rating usually significantly higher than those of other fuses | |
| Fuses to BS 1362 | 6.0 | – |
| Fuses to BS 3036 | According to category of duty marked on the fuse link, as follows: | |
| | S1 1.0) for all current ratings except 100A | |
| | S2 2.0) | |
| | S4 4.0 for current ratings from 30A up to and including 100A | |
| Miniature circuit breakers to BS 3871 | According to category of duty marked on the circuit breaker, as follows: | |
| | M1 1.0 | 0.85 to 0.9 |
| | M1.5 1.5 | 0.8 to 0.85 |
| | M2 2.0 | 0.75 to 0.8 |
| | M3 3.0 | 0.75 to 0.8 |
| | M4 4.0 | 0.75 to 0.8 |
| | M6 6.0 | 0.75 to 0.8 |
| | M9 9.0 | 0.55 to 0.6 |
| Circuit breakers to BS 4752 | Marked with the rated short circuit breaking current | |
| Factory-built assemblies to BS 5486 | Equipment marked with an indication of the short circuit strength | |

necessity electrically to protect the cables between the origin and the main distribution point of the installation, this point, for the small installation such as the typical household installation being the consumer unit. The cables concerned are, of course, the so-called 'meter tails' but these must not be excessively long so as to reduce the prospective short circuit current at the input terminals of the consumer unit to such a value that the disconnection time of the supply undertaking's fuse became such that the service cables reached excessive temperatures. The supply undertaking may specify both the maximum length and the minimum cross-sectional area of these meter tails where the supply undertaking's cut-out fuse is providing short circuit protection.

*Regulation 434-4*, which as already stated, gives the key requirement as regards short circuit protection, allows a lower breaking capacity than the prospective short circuit current provided that on the supply side of the protective device being considered there is another protective device of such a type that its energy let-through is not greater than that which can be withstood by the first mentioned device and the conductors being protected.

The concept of energy let-through can be appreciated by considering the current flowing during a short circuit fault, as shown in Fig. 21. The dotted curve represents one half-cycle of the prospective fault current, the fault being initiated at 0. At time $t_1$ the current has provided sufficient energy to melt a fuse element or open the contacts of an mcb. The area $OAt_1$ thus represents the 'pre-arcing' or 'operating' energy. The arc produced by circuit interruption still has to be extinguished and the energy passing during this process is represented by area $t_1At_2$. The total area $0At_2$ represents the total energy let-through. Although both $t_1$ and $t_2$ are shown within the first half-cycle of fault, $t_2$ may not occur before the current zero ($t_0$) if a zero point extinguishing breaker is used and both $t_1$ and $t_2$ could be delayed until subsequent half-cycles for comparatively low values of fault current.

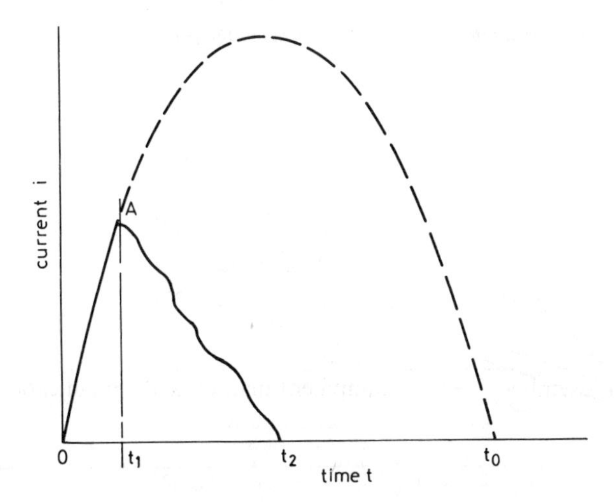


**Fig. 21** *Illustrating cut-off and energy let-through of overcurrent protective device*

Under heavy fault conditions, the arcing $I^2t$ becomes a considerable part of the total $I^2t$ but a fuse or certain types of mcb can 'cut off' or limit the current as shown in Fig. 21, so clearing the fault in less than half a cycle (i.e. in less than 0·01 s at 50Hz). In this case the device limits the $I^2t$ value and this is known as the 'energy let-through' for the device. For design purposes it then becomes necessary to use the manufacturer's $I^2t$ characteristics and cut-off characteristics, typical examples of which are shown in Figs. 22 and 23, respectively.

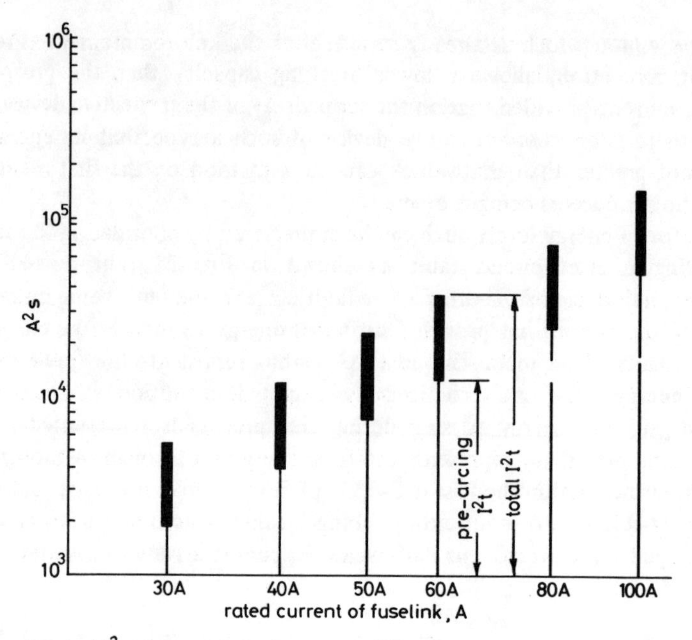


**Fig. 22** *Typical $I^2t$ characteristics for range of fuselinks (r.m.s. symmetrical current 33kA)*

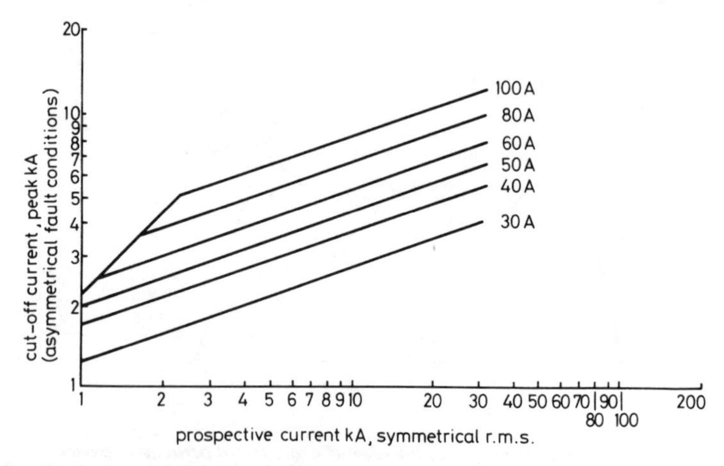


**Fig. 23** *Typical cut-off characteristics for a range of HBC fuselinks*

Fig. 24 shows the current traces for cut-off currents for a particular prospective current and in this very brief, and therefore incomplete description, it only remains to emphasise that the ability of a device to limit the fault energy applies only at fault currents greater than the relevant cut-off value.

At low fault prospective currents when the total operating time is longer than 0·01 s (one half-cycle at 50Hz) the arcing $I^2t$ is negligibly small compared with the pre-arcing $I^2t$ and can, for all practical purposes, be ignored. In this case one can use the time/current characteristics such as those included in Appendix 8 of the Wiring Regulations for design purposes.

Thus, as already stated, the energy let-through determined for the device on the supply side of the device actually under consideration must not be greater than the latter can withstand and it must also not be greater than the product $k^2S^2$ for the conductors of the circuit being protected, as indicated in *Regulation 434-6*.

It will be seen that the Note to *Regulation 434-6* refers to asymmetry, and this aspect is of particular importance when considering the suitability of an overcurrent protective device situated close to the output terminals of a generator or transformer. For instance, a 1000 kVA three-phase transformer having a percentage impedance of 5% will have a nominal fault level of 20 MVA at the low voltage (secondary) terminals. If the rated secondary voltage is 415 V this means that a fault current of 28 000 r.m.s. amperes or approximately 65 000 asymmetrical amperes will flow in a short circuit close to those terminals. These values are reduced the further away the fault is from the generator or transformer and, in general, for an installation fed from the public supply network, the design of the short circuit protection within that installation can be based on the r.m.s. prospective current.

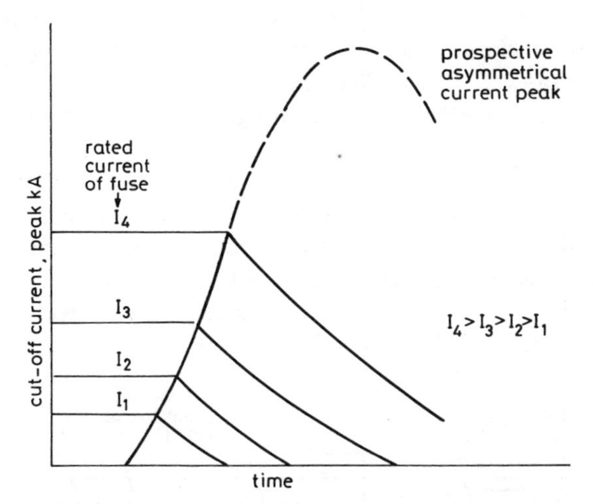


**Fig. 24** *Cut-off currents for a range of HBC fuses, for a particular prospective current*

An asymmetrical short circuit for example, occurs if at the instant of short circuit the electromotive force (EMF) wave of a generator is at, or near, its zero value and the peak value of the first asymmetrical short circuit current wave can then be approximately twice the peak value of the sustained symmetrical short-circuit current. The time taken for the current to reach the symmetrical condition depends on the time constant of the circuit, i.e. the ratio of its inductance to its resistance. The heating effect (and the electromagnetic stresses) are therefore much greater under the asymmetrical short circuit condition and the Note to *Regulation 434-6* indicates that where that condition is likely to be encountered the energy let-through of the short circuit protective device must again be less than the product $k^2 S^2$ for the conductors being protected.

*Regulation 473-5* indicates that, in general, all conductors must be protected against short circuit by a protective device at their origin. The only exceptions are recognised in *Regulations 473-6* and *473-8* where the former allows the protective device to be placed elsewhere than at the origin provided that the length of conductor left electrically unprotected is very short, that the risk of short circuit is very slight and that no danger could arise even if the conductors concerned were unexpectedly subjected to short circuit current. The situation envisaged by *Regulation 473-6* can occur in interconnections of switchgear but even in this case, considerable care must be exercised.

Where the electrically unprotected conductors are of much lower current-carrying capacity than that of the conductors from which they are supplied, they can in the extreme case be regarded as expendable; but as their failure on a short circuit could introduce a risk of fire, or danger to persons from molten metal or flash burns, or feedback of the fault onto the larger conductors such short runs therefore need to be designed so as to minimise the risk of short circuit and so that the effects of any unexpected fault are properly contained. Some guidance on this subject is given in BS 5486 for factory-built assemblies of switchgear, but where the conductors concerned are not enclosed in such an assembly the mechanical protection required to satisfy items (ii) and (iii) of *Regulation 473-6* may be provided by metallic conduit or trunking or similar enclosures in rigid p.v.c. If the enclosure is metallic it has to be earthed.

*Regulation 473-7* indicates that the short circuit protective device may also be placed at a point other than where the reduction in cross-sectional area occurs provided that the conductor having the smaller cross-sectional area situated between the two points is adequately protected by a short circuit protective device on the supply side of the point of reduction, 'adequately protected' meaning that the thermal requirement of *Regulation 434-6* is met.

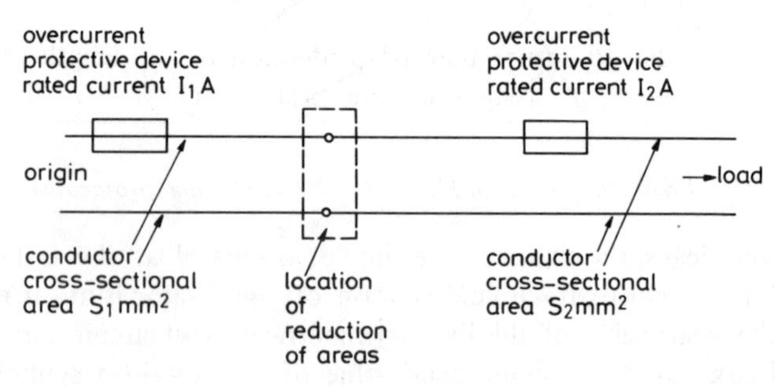


**Fig. 25** *Change of cross-sectional areas*

Fig. 25 shows this case in diagrammatic form. Two cables having different cross-sectional areas, $S_1$ and $S_2$ are in series, $S_1$ being the larger. At the supply end of this arrangement there is a short circuit protective device having a rated current of $I_1$ A and the other, on the load side of the point of reduction, having a rated current of $I_2$ A, $I_2$ being less than $I_1$.

Fig. 26 shows the time/current characteristics for the two protective devices, assuming they are fuses, together with the adiabatic lines for the two cross-sectional areas involved. The current $I_{F1}$ corresponds to the point of intersection of the time/current characteristic for the higher rated fuse and the adiabatic line for the larger conductor size. It is therefore the minimum fault current which could have been tolerated had there been no reduction in the conductor size. The current $I_{F2}$ corresponds to the point of intersection of the time/current characteristic for the lower rated fuse and the adiabatic line for the smaller conductor size and is the minimum fault current which could have been tolerated on the load side of that fuse.

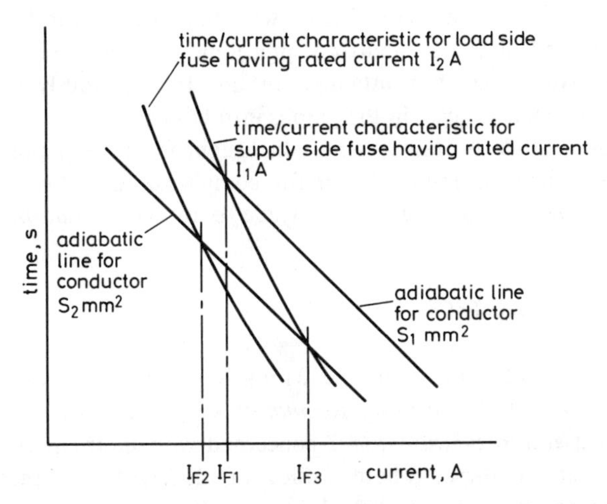


**Fig. 26** ***Minimum fault current with reduction of conductor cross-sectional area***
Both axes are logarithmic

However, in order to ensure thermal protection of the length of the smaller conductor size between the two fuses, the circuit impedance to the lower rated fuse must be such that the fault current is at least that corresponding to the point of intersection of the time/current characteristic of the *higher* rated fuse and the adiabatic line for the *smaller* conductor size, i.e. $I_{F3}$ in Fig. 26.

It follows from the foregoing that the situation could occur where the short circuit protective device preceding, on the supply side, the point of reduction, itself provides adequate protection of the 'reduced' conductor up to its termination or to the next load-side short circuit protective device. In such a situation, *Regulation 473-6* effectively permits the device at the point of reduction otherwise demanded by *Regulation 473-5* to be omitted.

This application of *Regulation 473-6* might be encountered, for example, in busbar tap-off connections or where a section of conductor has been increased due to grouping or other derating factors but is then reduced where it becomes separate from the group of conductors or is no longer subject to the other derating factors.

Returning to the graphs of Figs. 14 to 21, these are not applicable where the overcurrent protective device is intended to protect against overload currents only and when assessing the information given in them it must be remembered that if the overcurrent protective device is intended to protect against short circuit only, its rated current (according to *Regulation 434-1* as has already been mentioned) may be greater than the current-carrying capacity of the conductors it is protecting. If, on the other hand, the overcurrent protective device is intended to protect against both overload and short circuit currents, its rated current must not exceed the current-carrying capacity of the conductors it is protecting, this being demanded by *Regulation 432-2*.

Bearing in mind that the cross-sectional area of the conductors concerned will depend on the ambient temperature, the installation method used, grouping with conductors of other circuits, compliance with the voltage drop limitation, and in the case mentioned in the immediately preceding paragraph on the type of overcurrent protective device it is intended to use, it is obviously impracticable to translate the information given in the graphs into tabular form.

It is also impossible to draw general conclusions from the graphs and it is necessary to check individual circuit design for compliance with the thermal requirements of *Regulation 434-6*, unless it is possible to invoke *Regulation 434-5*.

## 6.5 Conductors in parallel

Two regulations in Chapter 43 of the Wiring Regulations are concerned with the protection of conductors in parallel, *Regulation 433-3* dealing with their protection against overload and *Regulation 434-7* concerned with short circuit protection. As regards the first of these, where cables of different cross-sectional area or disposition are connected in parallel, the current flowing through them depends on their respective impedances and will not necessarily divide in direct proportion to their respective cross-sectional areas.

Therefore, if two cables of different current-carrying capacity in parallel are intended to be protected by a common device having a rated current equal to the sum of the two current-carrying capacities, the cable having the lower capacity is likely to be insufficiently protected against overload. This particular point needs special attention in cases where one cable is installed to reinforce another.

As regards short circuit protection, it must be borne in mind that where conductors are paralleled, if a short circuit occurs it may involve only one set of the paralleled conductors. The distribution of the fault current in those conductors will depend on where the short circuit occurs relative to the origin of the circuit and it can be shown that the further the fault is from the origin the more onerous the thermal requirement of *Regulation 434-6* becomes. The analysis of the problem is basically the same as that developed in Appendix D of this Commentary for ring circuits except that instead of using earth fault loop impedances in one's calculations it is now necessary to use phase to neutral impedances.

When the paralleled conductors have the same cross-sectional areas (i.e. when $d$ = 1 in Fig. 27) the maximum impedance to the short circuit is attained when the short circuit occurs at the load end of the parallel circuit and the simplest design approach in order to meet the thermal requirements of *Regulation 434-6* is to see that the minimum short circuit current ensures disconnection in a faster time than that corresponding, as shown in Fig. 44 of Appendix D, to the point of intersection of the time/current characteristic of the device and the adiabatic line for the conductor cross-sectional area concerned. In other words, see that the short circuit device of the circuit gives satisfactory protection for each set of the paralleled conductors as if that set were carrying the entire fault current.

If the paralleled conductors do not have the same cross-sectional areas, the problem becomes slightly more complex. It can, for instance, be shown that the maximum impedance to the fault no longer corresponds to the fault being located at the load end, and it may then become necessary to carry out a more precise calculation. On the purely theoretical basis that the fault is a 'dead' short circuit and the resultant fault current divides in exactly the same proportion as the relevant impedances Fig. 27 shows the effect on the overall impedance presented to a fault in one of two paralleled single-phase cables.

Fig. 27 has been developed in the following manner:

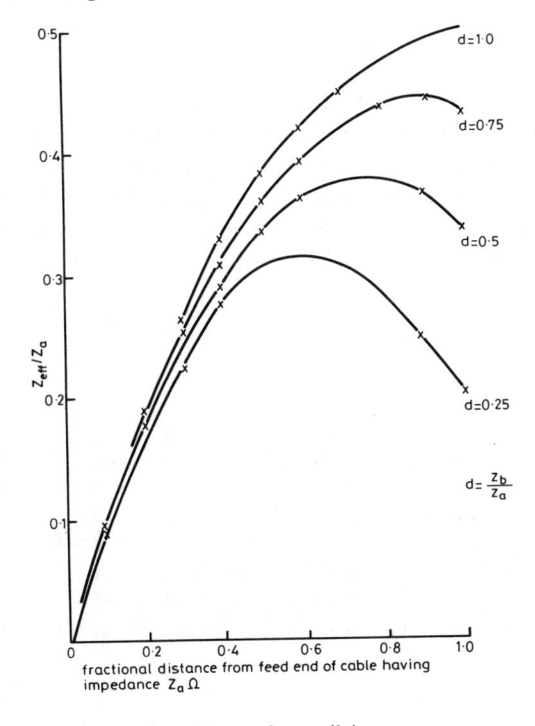


**Fig. 27** *Effective impedance of conductors in parallel*

Fig. 28 shows two paralleled cables where $Z_a$ is the phase plus neutral impedance of cable A and $Z_b$ is that of cable B. A dead short circuit has occurred in cable A at the fractional distance $x_a$ from the feed end. The equivalent circuit is therefore as shown in Fig. 29.

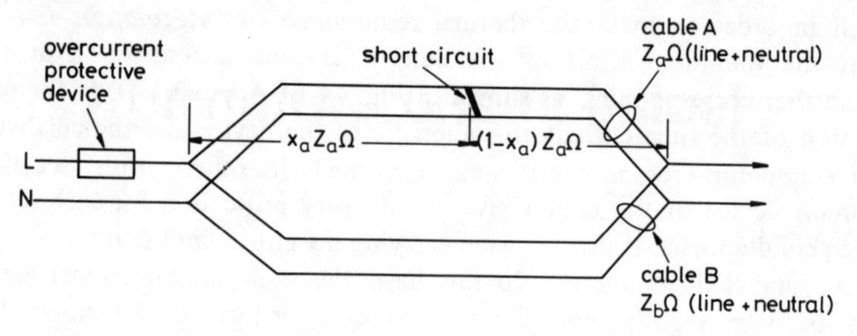


**Fig. 28** *Parallel cables and short circuit*

If the ratio of the impedances of the cables $Z_b/Z_a$ is denoted by '*d*' (and for the purposes of this analysis, the cables are designated '*a*' and '*b*' so that this ratio is less than unity), it can be shown that

$$\frac{Z_{eff}}{Z_a} = \frac{X_a(1 - X_a + d)}{(1 + d)}$$

where $Z_{eff}$ is the effective or equivalent impedance.

It is this equation which has been used to plot the family of curves shown in Fig. 27. The maximum effective impedance, giving rise to the minimum short circuit current, occurs when the short circuit takes place in the cable having the higher impedance $Z_a$ and, as would be expected, is always less than the maximum effective impedance when the paralleled cables are of the same size.

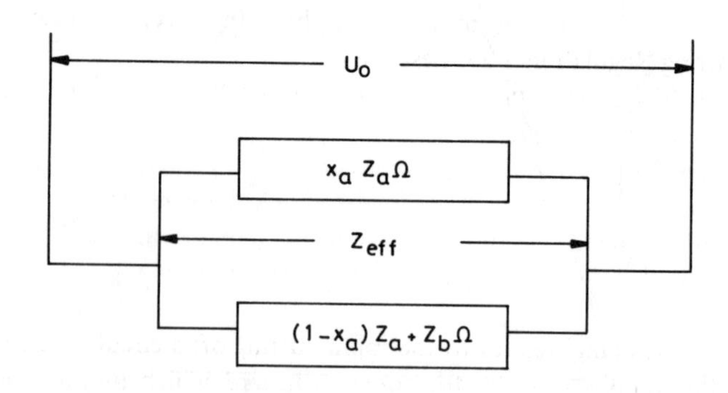


**Fig. 29** *Equivalent circuit diagram*

Chapter 7

# Isolation and switching

Remembering that Part 4 of the Wiring Regulations deals with measures of protection for safety, Chapter 46 is confined to isolation and switching from that aspect, Section 476 deals with the application of these protective measures and Section 537 prescribes the requirements for isolators and switching devices.

Functional switching, which may be defined as switching for convenience in the normal operation of equipment, therefore, is not treated in Chapter 46 of the Wiring Regulations, although functional switching devices may be used for isolation and safety switching provided that they comply with the relevant requirements of Chapter 46 and the two companion sections referred to above. Those sections, however, do include requirements concerning functional switching.

As a result of international discussions, three distinct and separate types of isolation and switching are now specified and this approach is new to the Wiring Regulations. The treatment of these aspects of isolation and switching is one example of the way in which the Wiring Regulations have aimed to be more specific in their dealing with requirements for industrial and other non-domestic installations. An understanding of the meanings of these three types is essential for the user of the Wiring Regulations and these are as follows.

'Isolation' is limited strictly to cutting off an installation or part of an installation from all sources of energy and this measure, where used alone is intended to enable skilled persons to carry out work on, or in the vicinity of, parts which would otherwise be live. 'Switching for mechanical maintenance' relates to the electrical inactivation of electrically powered equipment, to enable non-electrical work to be carried out on the equipment by persons who may not be electrically skilled.

'Emergency switching' relates to the rapid cutting-off a circuit or item of equipment from the supply in order to remove a hazard which may be electrical or mechanical. The requirements concerning this measure differ accordingly and under some circumstances may even involve the energising of devices such as electromagnet brakes. Before commenting separately on the three functions or types of isolation and switching, *Regulation 476-1* makes it quite clear that means of isolation must *always* be provided whereas a particular installation may be of such a character that neither switching off for mechanical maintenance nor emergency switching is required. *Regulation 476-1* also states that a device can be used to perform more than one of the three functions, (and this is frequently the case in practice), provided that the device meets all the requirements for the functions it is intended to carry out. The device intended to meet the other requirements for switching as prescribed in *Regulations 476-15* and *476-16* may also serve to fulfil one or more of the above functions.

## 7.1 Isolation

As indicated by *Regulation 537-3*, an essential feature of an isolating device is that it must provide a specified distance of contact separation and specified creepage and clearance distances and therefore not all switches are suitable for use as isolators, a particular example of a switch not providing adequate contact separation being the microgap switch. If there is any doubt as to the suitability of a particular type of switch or other device for use as an isolator the manufacturer should be consulted.

There are two further essential features, the first of these being that either the position of the contacts is visible or a reliable means of indication is given, as demanded by *Regulation 537-5*. The second feature is a means to prevent inadvertent operation as demanded by *Regulation 537-7*

Devices which are suitable for the purpose of isolation include

(*a*) isolators (disconnectors)
(*b*) switch-disconnectors (isolating switches)
(*c*) plugs and socket outlets
(*d*) fuselinks
(*e*) links
(*f*) circuit breakers having the required contact separation.

An isolator, by definition, has to be capable of opening and closing a circuit when either negligible current is broken or made, or when no significant change in the voltage across the terminals of each of the poles of the isolator occurs. It is also capable of carrying currents under normal circuit conditions and carrying, for a specified time, currents under abnormal conditions such as those of short circuit.

*Regulation 461-1* requires that in general all live conductors must be disconnected for isolation to be effective. A neutral conductor is, by definition, also a live conductor and this regulation, by cross-referring to *Regulation 460-2*, indicates that the neutral conductor need not be isolated in a TN-S system. A neutral link is often provided for testing purposes but it is not necessary to provide one for isolation as such. In no circumstances may a means of isolation be connected in a protective conductor.

The reference in *Regulation 460-2* to Section 554 is needed because, for high voltage discharge lighting installations where an autotransformer is used on an a.c. two-wire circuit in a TN or TT system, both poles of the supply are required by *Regulation 554-6* to be provided with means of isolation. *Regulation 460-2* also indicates that the one case where the means of isolation shall *not* disconnect the neutral is that of the PEN conductor in a TN-C system, i.e. the combined neutral earth conductor in an earthed concentric wiring system.

Returning to *Regulation 461-1*, this also indicates that a single means of isolation may be used for a group of circuits and this facility applies even to motor circuits which are the subject of *Regulation 476-5*. The actual number of isolating devices used in an installation will, of course, depend on the extent and nature of the installation.

*Regulation 461-5* requires that all devices used for isolation purposes shall be clearly identifiable, and this aspect of identification is of particular importance where one isolator serves a number of circuits.

For the small installation, one means of isolation may be sufficient, but for other installations it seems obvious that there must be more as it would be at least highly inconvenient to shut down a whole installation in order, for instance, to effect repairs to one item of equipment in a safe manner. Because of this aspect of possible inconvenience, there would be a risk of persons attempting to work on live equipment without first isolating the supply to that equipment and to avoid this risk an adequate number of isolating devices should be provided. Careful consideration should also be given to the location of these devices because if they are not suitably placed in relation to the circuits or equipment with which they are associated there is a similar risk that they would not be used.

In many cases, of course, the means of isolation for individual circuits or sections of an installation will be the same device which serves another of the functions covered by Chapter 46 of the Wiring Regulations. For an installation serving more than one building, if the separated buildings are treated as separate installations, *Regulation 476-1* requires a separate means of isolation for each building. Although the isolator (and, incidentally, any device used for mechanical switching maintenance) may be located remote from the building concerned, the need for disconnection when access to another building for the purpose might be impeded or cause undue delay should be borne in mind, in the spirit of *Regulation 314-1*.

As regards *Regulation 476-2*, it is pointed out that the Note to *Regulation 13-14* makes it clear that main switchgear or fusegear provided by the electricity supply undertaking may be used as the main means of isolation of the consumer's installation where this is permitted by the supply undertaking and the consumer is prepared to request the supply undertaking to disconnect and reconnect whenever work needs to be done or part of the installation between its origin and the first main distribution point, or in the smaller installation, between its origin and the switch provided for compliance with *Regulation 476-15*.

*Regulations 461-2* and *476-4* are two of the most important concerning isolation, and are intended to ensure that in practice having, for instance, isolated a circuit to enable someone to effect repairs on an item of equipment in that circuit, every effort is made to prevent the reenergising of the circuit thereby placing that person at risk. Precautions for compliance with these regulations may take the form of padlocking, location of the device in a lockable room or other enclosure and/or short circuiting and earthing of the isolated live parts. Furthermore, as isolation is an operation required to be carried out by skilled persons for the purpose of maintenance, repairs and the like, it is not necessary for the isolating device to be readily accessible, if isolation is its only function. The user of the installation, unless a skilled person, is not expected to operate such isolating devices because switches are provided for the normal functioning of the installation.

*Regulation 476-4* indicates that there is also no objection to the means of isolation being remote from the equipment concerned provided that either adequate precautions are taken to prevent inadvertent reclosure of the device such as those just detailed, or an additional isolating device is placed adjacent to the equipment.

*Regulation 461-3* requires that a warning notice is used whenever an item of equipment contains live parts connected to more than one supply, unless an interlocking arrangement is used. This requirement applies whether the various supplies are obtained from the same phase or from different phases or from different sources entirely, the various supplies being considered to be separate if they do not share a common isolating device. A similar provision exists in British Standards for equipment. For example, BS 3456, in dealing with requirements for stationary heating appliances for multiple supply, prescribes that a warning notice must be incorporated in these appliances stating that all the supply circuits must be isolated before obtaining access to terminal devices.

BS 3456, in common with a number of other equipment standards, also requires, for stationary equipment not provided with a non-detachable flexible cable or cord intended to be connected to a plug or with other means for disconnection from the supply having a contact separation of at least 3mm in all poles, that the instruction sheet accompanying the equipment must state that such means of disconnection has to be incorporated in the fixed wiring.

## 7.2 Switching-off for mechanical maintenance

Mechanical maintenance includes, for example, the cleaning of equipment, replacement of lamps and the maintenance of electrically driven machines. For the last mentioned there may be mechanical means of inactivating the drive, such as a clutch or the disengagement of a belt; in such cases electrical switching for mechanical maintenance would not be essential but, where provided, it should still comply with the requirements of Sections 462, 476 and 537.

The major difference between switching for mechanical maintenance and isolation is that the former function is likely to be carried out by other than electrically skilled persons and therefore no access to live parts is permitted, as stated in *Regulation 476-8*. The device, if a switch, should be capable of cutting off full load current but it is not necessary to switch all the live poles of the supply (*Regulation 537-11*). In general this means that it is not necessary to switch the neutral.

*Regulation 537-8* indicates that the devices should be inserted in the main supply circuit but accepts that in certain circumstances they may be fitted in the control circuits of equipments.

In the design of control circuits, consideration should be given to the use of a reduced voltage, particularly where these control circuits are extended remotely from the equipment circuit. A 110 V supply is frequently used for this purpose but must not be confused with the 110 V centre-tap-earthed system covered by *Regulations 471-28* to *471-33* and commented on in Chapter 4 of this Commentary.

The precise arrangements used for this function depends on the nature of the maintenance but the devices used must comply with *Regulations 537-8* to *537-11*. It will be noted that such devices or their control switches must be manually operated as prescribed in *Regulation 537-9* and shall be selected and/or installed so as to prevent unintentional reclosure as prescribed in *Regulation 537-10*.

*Regulation 462-2* requires that devices for switching off for mechanical maintenance are to be suitably placed, readily identifiable and convenient for their intended use. However, for infrequent cleaning of equipment, the means of switching can be remote, provided, as already mentioned, that precautions are taken against reactivation. On the other hand, where there is a frequent need for persons to clean or adjust machines during the normal use of those machines, the devices for switching-off for mechanical maintenance should be local to the machines and readily accessible to those persons. Here too there is a need to take precautions against reactivation and these precautions should also take into account the design of the circuit so that, for instance, pressing any 'start' button will not reactivate the machinery until the switching off device has been released.

Devices which are suitable for switching off for mechanical maintenance include

(*a*) switches
(*b*) circuit breakers
(*c*) control switches operating contactors
(*d*) plugs and socket outlets.

Switching-off for mechanical maintenance is required only where mechanical hazards are foreseen, and *Regulation 476-7* recognises this by demanding that a device for this function be provided for every motor circuit or for a circuit supplying electromagnetic equipment for operations from which mechanical accidents could arise. However, care should be exercised where opening of the circuit could cause danger, for example, in supply circuits for lifting magnets. *Regulation 476-7* also demands switching off for mechanical maintenance for circuits supplying equipment having electrically heated surfaces which can be touched.

## 7.3 Emergency switching

The essential aspect of the measure known as emergency switching, as indicated in *Regulation 463-1*, is that it is intended to disconnect the circuit concerned as rapidly as possible to prevent or remove a hazard. All parts of an installation, and the complete installation, must of course incorporate means of switching off to remove or prevent a hazard but this is not emergency switching unless the disconnection needs to be immediate, and these other means are considered later in this Chapter of the Commentary.

The devices for the purpose of emergency switching include
(*a*) switches in the main circuit
(*b*) push-buttons and the like in a control or auxiliary circuit,
and the requirements for these devices are prescribed in *Regulations 537-12* to *537-17*, the last being concerned with the fireman's emergency switch associated with a discharge lighting installation. The fireman's emergency switch is a good example of a switch which will not only perform the function of emergency switching but also that of isolation.

Particular attention should be paid to the requirements prescribed in *Regulation 537-16*. Emergency switching devices should preferably be manually operated as indicated by *Regulation 537-14* and must be readily available in the immediate vicinity of the equipment likely to cause a hazard as required by *Regulation 537-15* but, if considered appropriate, also at any additional remote position from which another person can quickly remove the hazard.

## 7.4 Other requirements for switching

The remaining regulations in Section 476 which have been grouped under the heading 'Other requirements for switching' recognise the fact that although there are many applications where emergency switching, as defined earlier in this present chapter, is not required, it is nevertheless essential that there should be a safe means of switching off available to the user of the installation.

Thus *Regulation 476-15* requires a main switch or circuit-breaker to be provided for an installation, *Regulation 476-16* requires the circuits within that installation to have a means of interruption (circuits may be grouped and switched by one device), and *Regulation 476-17* requires each appliance and luminaire to be controlled by a readily accessible switch, unless it is connected to the supply by means of a plug and socket outlet. This exemption nevertheless suggests that such a plug and socket outlet are themselves readily accessible to the user.

*Regulation 476-18* allows the use of a switch mounted on the appliance or luminaire (or integral with the appliance or luminaire) provided that the switch is primarily intended to serve as a means of switching off for mechanical maintenance and therefore has to comply with *Regulation 476-8.*

*Regulation 476-20* concerning switches controlling fixed or stationary household cooking appliances also appeared in the Fourteenth Edition and although in general it has not caused practical difficulties, the limitation of the distance from the switch to the appliance has, in the past, led to the switch being so positioned that the user, in order to operate it, has had to lean across the cooker. Obviously, in the case of a cooker fire, the user would then be exposed to some danger and this point should be taken into account when positioning cooker switches. If the designer or contractor knows the type and make of cooker it is intended to use it seem that a sensible approach would be to install the cooker switch to one side of the cooker, taking the fullest possible advantage of the 2m the regulation permits. This distance is the physical distance between the cooker switch and the cooking appliance and does not refer to the length of the connecting cable. The same reasoning applies to a socket outlet positioned over, or near, a cooker hob.

Chapter 8

# Common rules for the selection and erection of equipment

## 8.1 Compliance with standards

*Regulation 511-1* requires all items of equipment to comply with the relevant requirements of the current edition of the applicable British Standard and as Appendix 1 of the Wiring Regulations shows, cross-reference is made to a large number of British Standards. The aim has always been to have, wherever possible, complete compatibility between the Regulations and the requirements prescribed in the British Standards, many of which, in their turn, cross-refer to the Wiring Regulations.

In fact, the Terms of Reference of the Wiring Regulations Committee state that in drafting any requirements in regard to materials and the construction of equipment, it is their duty:

(*a*) to require the compliance of appliances and materials with such requirements of the appropriate British Standards as refer to safety in operation, and

(*b*) to formulate statements of conditions necessary to the safety of appliances permitted in the Regulations but not covered by British Standards, and co-operate with the BSI in the preparation of the British Standards required.

It will be noted that item (*a*) is specifically limited to compliance with the requirements of the equipment standards *as refer to safety in operation*. At the time when the Committee's Terms of Reference were promulgated a British Standard frequently included both safety and performance requirements but the IEE Wiring Regulations were limited, as they still are, to providing safety, and so it was both logical and necessary to limit the demand for compliance with the relevant British Standards to only the safety requirements embodied in those standards.

However, for some years it has been the policy of BSI to segregate safety requirements from those concerning performance. For example, BS 4934 covers the safety requirements for electric circulating fans while the performance requirements are prescribed in BS 5060.

As another example, the comprehensive BS 3456 series of British Standards covers the safety requirements for household electrical appliances, but there is no British Standard prescribing performance requirements for such products, although BS 3999 gives standard methods of measuring performance which are intended to be used in association with, for instance, consumer labelling schemes.

It must be emphasised that the Wiring Regulations demand compliance with the relevant British Standard, not that the product concerned has to have the certification mark of the appropriate certifying organisation – one of those named in Note 4 to *Regulation 511-1*.

Whether one uses a product having, for instance, the BS Safety Mark, or one from a manufacturer who considers it unnecessary to obtain that Mark, is a matter of personal choice and judgement. The British Standards themselves as a general rule have no requirement that the product dealt with has to be submitted to the appropriate certifying body; for instance, BS 6004 which specifies requirements and dimensions for p.v.c.-insulated cables merely draws attention to the certification services offered by the British Approvals Service for Electric Cables (BASEC).

The Preface to the Wiring Regulations makes reference to the use of equipment complying with a foreign standard. If it is intended to use imported equipment claimed to be in accordance with a particular foreign standard it is first necessary to check if that standard implements a CENELEC Harmonization Document and, if it does, examination of the HD (obtainable from BSI) will indicate whether the foreign standard includes any departure from the HD, i.e. a national deviation, which would adversely affect the ability of the equipment to comply with the Wiring Regulations.

If the foreign standard is directly based on an IEC Standard and not a CENELEC HD, there may be no such ready method of checking whether the foreign and the corresponding British Standard are identical. The only way, in this case, is a clause-by-clause comparison.

Mention has been made earlier that CENELEC also has the opportunity, when considered appropriate, of producing European Standards. When this method is adopted no national deviations are permitted so that all the national standards implementing a particular European Standard are identical.

The need to have compatability between national and international wiring rules and between those and national and international equipment standards is of considerable importance to equipment manufacturers because, if compatability is not obtained, the situation can arise where the manufacturer has to meet differing or additional requirements in the countries in which he wishes to sell. Although those differences or additions may appear to be trivial, they could involve the manufacturer in considerable tooling costs and perhaps expensive changes in manufacturing methods without any real advantage either in safety levels or performance.

The purpose of Note 1 to *Regulation 511-1* is to take account of the case where, during the preparatory stages of an installation, a significant change occurs in the requirements of an applicable British Standard. The parties to the particular contract concerned should then consider whether it is necessary to take account of that change, i.e. to agree upon the date of the British Standard to be observed.

It may well happen that a British Standard is revised or amended in the interval between successive periodical inspections and testing of an installation, the revision or amendment being of such a nature that at the later periodical inspection it could no longer be claimed that the installation complied with *Regulation 511-1*. This fact should be noted on the inspection certificate and it is then for the parties concerned to decide on the course of action to be taken.

Note 3 to *Regulation 511-1* has the effect of allowing the use of equipment not covered by a British Standard, but only if that equipment gives an adequate level of safety. This Note therefore allows, for example, the use of industrial plugs and socket outlets which have features which mean that they are not covered by the British Standard generally accepted as the 'applicable' one, namely BS 4343. For instance, they may have the pin sizes and centres of plugs and socket outlets to BS 1363 and have configurations or enclosures not covered by that standard but which are more akin to BS 4343. In such cases it is nevertheless expected that these accessories would meet requirements extracted from both the British Standards mentioned which are considered to be relevant, in order to give the level of safety required.

## 8.2 Operational conditions and external influences

Section 512 lists the operational requirements for which equipment must be suitable and *Regulations 512-1, 512-2* (ii) and *512-3*, dealing with voltage, current and frequency, respectively and which must be complied with, are directly related to *Regulation 313-1* which, it will be recalled, requires these aspects of the supply to be ascertained as part of the first action to be taken in installation design. If there is any doubt as to the suitability of particular switches or circuit breakers for use with inductive or capacitive equipment the advice of the switch or circuit breaker manufacturer should be sought.

The external influences have already been mentioned in commenting on Chapter 32 and here, in relation to the equipment used in the installation, it is only necessary to point out that external influences must always be assessed as they will exist at the point in the completed installation at which the equipment is to be installed. As a simple but important example of this, the ambient temperature in household premises is unlikely to exceed 30°C if one is considering the premises generally, but may be appreciably higher in certain locations where equipment is installed such as in airing cupboards and in the vicinity of heating appliances. It is therefore necessary to verify that accessories and other equipment to be installed in such locations are suitable for the local ambient temperature expected while carrying their rated currents.

## 8.3. Accessibility

Section 513 - Accessibility comprises only one regulation; this is couched in very general terms and there are relatively few regulations in the Wiring Regulations other than in Section 526 which give more detailed or specific requirements as regards this aspect. While this Section 513 applies to all types of installation, it is of particular importance in industrial installations and a number of individual regulations in the Electricity (Factories Act) Special Regulations 1908 and 1944 are concerned with ensuring safe working space and means of access to equipment that has to be worked or given attention or, as another example, with the accessibility of parts of switchboards requiring to be adjusted or handled. The Memorandum on the Electricity Regulations gives a great deal of guidance on this question of accessibility.

## 8.4 Identification and notices

The next section of Chapter 51, namely Section 514, deals with identification and notices and some of the regulations in this Section embody requirements which are new to the Wiring Regulations.

*Regulation 514-1* introduces the requirement that where the operation of switchgear and controlgear cannot be seen by the operator and this may create a danger, a suitable indicator to BS 4099 has to be visible to the operator, that British Standard dealing with the colours which should be used for indicator lights, push buttons, annunciators and digital readouts.

*Regulation 514-3* introduces also for the first time in the Wiring Regulations a requirement for the provision of diagrams, charts or tables describing the installation. This requirement is based on international agreements and is needed because the details of the installation demanded here are necessary for verifying compliance with the Wiring Regulations (See *Regulation 611-2*). For example, details of the types of protective device used are essential for the verification of compliance with the requirements for protection against electric shock and overcurrent.

*Regulation 514-3* does not give a detailed prescription as to the form the charts or tables should take but Table 16 is one suggested version of a schedule which could be used for the smaller installation. Fig. 30 illustrates a typical schematic diagram for a small industrial or commercial installation on which is indicated the information required for compliance with *Regulation 514-3*. Should the ownership of the premises concerned change hands it is very important that the new owners have the fullest possible information concerning the electrical installation and it is essential that diagrams, charts, tables and schedules are kept up-to-date. Such items, in any event, are invaluable aids in the maintenance of an installation. In the more complex installation, the information demanded by *Regulation 514-3* may sensibly be accompanied by specific mention of circuits where it may not be immediately apparent what protective measures have been used. For instance, the protective conductors used for earth free local equipotential bonding are colour coded green-and-yellow in exactly the same manner as other protective conductors. Should they be mistaken for circuit protective conductors by someone unfamiliar with the

**Table 16: Schedule of installation at . . . . . . . . . . . . . . . . . . . . , as prescribed in the IEE 'Regulations for electrical installations'**

| Type of circuit | Points served | Phase Conductor $mm^2$ | Protective Conductor $mm^2$ | Type of wiring | Protective and switching devices |
|---|---|---|---|---|---|
| Ring final circuit | General purpose 13A socket outlets | 2·5 | 1.5 | P.V.C.-insulated and sheathed, flat twin and earth | 30A miniature circuit breaker; local switches and plugs and socket outlets |
| No. 1 lighting circuit | Downstairs fixed lighting | 1.5 | 1.0 | " | 5A m.c.b; local switches |
| No. 2 lighting circuit | Upstairs fixed lighting | 1.5 | 1.0 | " | 5A m.c.b; local switches |
| Cooker circuit | Cooker | 6.0 | 2.5 | " | 45A m.c.b; cooker control unit |
| Immersion heater circuit | Immersion heater | 2.5 | 1.5 | " | 15A m.c.b; local switched connector box |

installation who then 'rectifies' what appears to be an omission by bonding the conductors to the main earthing terminal in order to earth them or by connecting them to exposed or extraneous conductive parts which are already connected to the main earthing terminal, the protective measure is then destroyed and a possibly dangerous situation could arise. Similar comments can be made of circuits where the protective measure is protection by electrical separation where again the measure depends on there being no direct connection with Earth but where there may again be equipotential bonding conductors.

*Regulation 514-4* has the main purpose of safeguarding the skilled person who is carrying out work on live parts intended to be isolated and in general terms the regulation applies only to three-phase circuits. Although the regulation is concerned with means of access to live parts, this does not mean that it is necessary to have an external indication on items of equipment which are simultaneously accessible and connected to different phases; the warning notice can, for example, be placed inside the means of access to live parts.

The notice concerning periodic inspection and testing demanded by *Regulation 514-5* now incorporates for the first time provision for inserting the date of the last inspection and for the recommended date of the next inspection. The periods between inspection are only recommended because it would be impracticable to place mandatory limits on these, the recommended periods being given in Appendix 16 of the Wiring Regulations but, as mentioned earlier, in premises subject to licensing, the authority concerned may prescribe the frequency of periodic inspections.

As regards *Regulation 514-6* which details the text which must be placed near the main switch of any caravan claimed to be wired in accordance with the IEE Wiring Regulations, it will be seen that in item 1(*c*), the user of the caravan on arrival at the site should check that any residual current device in the mains supply to the caravan had been tested within the last month.

This is a new instruction and is directly related to the new regulation, *Regulation 471-44*, which requires that the supply to each caravan on a site shall be protected by a residual current device either individually or shared with up to five socket outlets supplying other caravans.

The notice prescribed in *Regulation 514-8* is also related to a new Regulation, (*Regulation 471-12*), which is an important departure from previous editions requiring socket outlet circuits rated at 32A or less or other points of utilisation supplying equipment by means of a flexible cable or cord having a similar current-carrying capacity to be protected by means of a residual current device if the equipment being supplied is used outside the zone created by compliance with *Regulation 413-2* and may be touched by a person in contact directly with the general mass of Earth.

## 8.5 Mutual detrimental influence

Compliance with *Regulation 512-6* will normally mean that the electrical installation

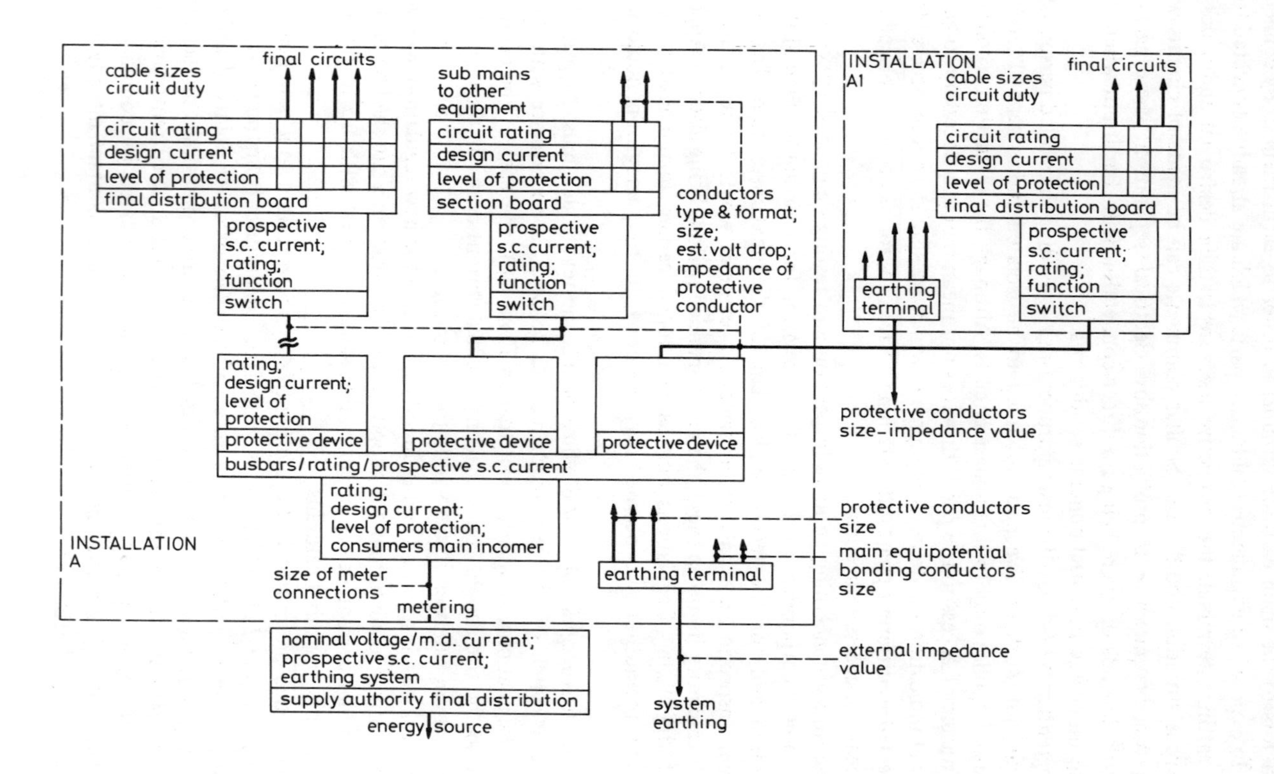


**Fig. 30** *Diagram to provide information as required in clause 514-3*

does not suffer any harmful influence from other types of installation but *Regulation 515-1* is also concerned that the electrical installation shall have no adverse effect on non-electrical services and one obvious precaution which should be taken, when practicable, is to arrange that the electrical installation is physically separate from the non-electrical services. Damage to the latter can occur, for example, from heating by leakage currents and by electrolysis so that due action needs to be taken to minimise these risks.

Chapter 9

# Cables, conductors and wiring materials

A great number of the regulations appearing in Chapter 52 will be immediately recognised as being from Part B of the Fourteenth Edition and this is so because discussions are still at a relatively early stage in the international work dealing with the subject matter of this Chapter.

What, however, is significantly different between the Fifteenth Edition and earlier editions is the basic approach which has been adopted by indicating in Section 523 'Environmental conditions' the reason for particular requirements or for the use of a particular type of cable by the placing of the regulations concerned under a specific sub-heading describing the pertinent environmental condition.

It will be seen that these environmental conditions align with some of the items in General Category 'A' of the classification of external influences as detailed in Appendix 6 of the Wiring Regulations and as work progresses internationally on the application of the classification Chapter 52 and particularly Section 523 may be required to be amended.

Appendix 10 of the Wiring Regulations gives in tabular form information on the application of commonly used cables taking account of the risk of mechanical damage and of corrosion. The tables, as indicated in the preamble of the Appendix, are not exhaustive and other external influences may have to be taken into account in selecting the type of cable to be used. The final choice of cable, in any event, may be determined by, for instance, a comparison between the maximum permissible operating temperatures (and hence of the current-carrying capacities) of types of cable equally acceptable in terms of their suitability as regards protection against mechanical damage or corrosion or both.

*Regulation 521-1* retains the limitation in the use of aluminium conductors which was in the Fourteenth Edition and therefore the minimum size of plain aluminium conductor which can be used is still 16 mm$^2$. Although Note 3 to *Regulation 521-1* indicates that care must be exercised when using mineral-insulated cables in discharge lighting circuits because of the voltage surges caused by the inductive nature of the ballast units, it must be remembered that such voltage surges, which may be of several kilovolts magnitude, can be caused by the switching of other types of inductive load.

In addition to ballast units, the main sources of trouble are contactor coils and three-phase star connected small power motors such as those used in heating and ventilating systems. Breakdown is most likely to occur in inductive circuits which are frequently switched and as mineral-insulated cables are particularly sensitive to these voltage surges, unless it is known that the equipment likely to cause trouble incorporates its own means of surge suppression, it is advisable to consult the manufacturers of the equipment if it is intended to supply them from mineral-insulated cables. Simple standard surge suppressors are available from the cable manufacturer and these should be fitted as close as possible to the terminals of the equipment likely to initiate the voltage surges but it is inadvisable to interfere with the internal wiring of the equipment.

When using high-pressure sodium and metal halide discharge lamps, the associated control gear is mounted separately from the lamp and this is also true in some mercury and fluorescent systems. In such cases, surges in the cables interconnecting the lamps and the control gear are inevitable and mineral-insulated cables should not be used although they can be used as the supply cables to the control gear.

## 9.1 Current-carrying capacity of cables

Section 522 deals with the operational conditions which must be taken into account when determining conductor cross-sectional areas and the first regulation in this Section, i.e. *Regulation 522-1*, gives the basic rule that the current-carrying capacity of a cable conductor must not be less than the maximum sustained current normally carried by that conductor.

The current-carrying capacity of a cable is determined by

(*a*) the conductor material
(*b*) the insulation material
(*c*) the ambient temperature
(*d*) the method of installation, including grouping with the cables of other circuits.

A cable having aluminium conductors has approximately only 80% the current-carrying capacity of a similar cable (i.e. having the same insulation material) with copper conductors of the same cross-sectional area, operated at the same ambient temperature and installed in the same manner.

The temperature at which the conductors of a cable may be allowed to operate continuously without damage to the cables, and for a reasonable service life, depends on the insulation material used in the construction of the cable. P.V.C.-insulated cables to BS 6004 are suitable for use where the conductor temperature under normal load conditions does not exceed 70°C, whereas 85°C rubber-insulated cables to BS 6007 are suitable for use, as their name suggests, where the conductor temperature under those conditions does not exceed 85°C.

Table 17 gives the permissible normal operating temperature for various types of insulating materials used in cable manufacture.

**Table 17: Maximum normal operating temperatures for conductors and cables**

| Type of insulation or sheath | Maximum normal operating temperature |
|---|---|
| 1 | 2 |
| | °C |
| 60°C rubber compounds (Types EI1, EM1, and heavy-duty to BS 6899) | 60 |
| 60°C oil-resisting and flame-retardant sheath compound (Type EM2, and heavy-duty O & FR to BS 6899) | 60 |
| PVC compound (Type TI1, TI2, TM1, TM2, 2 and 6 to BS 6746) | 70 |
| Impregnated paper[+] | 80 |
| 85°C rubber compounds (Types GP2, GP2, FR1, FR2, and OR1 to BS 6899) | 85 |
| PVC compound (Types 4 and 5 to BS 6746) | 85 |
| HOFR and heavy-duty HOFR compounds to BS 6899 | 85 |
| XLPE (to BS 5468 | 90 |
| HEPR (to BS 5469) | 90 |
| 105°C rubber (Type EI3 to BS 6899) | 105 |
| Mineral with 105°C terminations | 105* |
| 105°C rubber (Type EI2 to BS 6899) | 150 |
| Fluorocarbon compounds | 150 |
| Glass fibre with 155°C varnish | 155 |
| Glass fibre with 185°C varnish | 185 |
| Mineral with 185°C terminations | 185 |

+ Applicable only to cables of voltage rating 600/1000V

* For mineral-insulated cables, sheathed with p.v.c, the values for PVC (70°C) compound are applicable. Otherwise, the values shown for m.i. cables relate only to terminations; elsewhere the temperature of the cable should not exceed 250°C.

Note: Where the insulation and sheath are of different materials, the appropriate limits of temperature for both materials must be observed. For cables sheathed with materials not mentioned in the Table (e.g. metal), it may be assumed that the limits of temperature specified for the insulation will also provide a satisfactory sheath temperature.

In general, the maximum sustained current normally carried by a conductor is the design current ($I_B$ A) of the circuit concerned so that

$$I_B \leqslant I_Z \text{ A}$$

where $I_Z$ is the current-carrying capacity of the conductor. This equation has, of course, already been encountered in *Regulation 433-2*.

In some cases however, particular attention has to be paid to the operating characteristics of the current-using equipment fed by the cable being considered. For instance, the starting and accelerating currents of some types of motor are considerably in excess of the normal full load current. Particularly where a motor is operating intermittently and is subject to frequent stopping and starting, account has to be taken of the cumulative effects of the starting periods on the temperature rise of the cables carrying the starting current, as required by *Regulation 552-1*, and it may be necessary to seek advice from the motor manufacturer.

The operating temperature of a cable conductor is the sum of the ambient temperature and the temperature rise caused by the passage of the load current so that the current which can be carried by a conductor without exceeding the permitted maximum operating temperature, i.e. the current-carrying capacity of the cable, decreases as the ambient temperature increases.

It may well be that the route taken by a particular cable is such that the ambient temperature cannot be assumed to be constant over that route, and in such a case the highest ambient temperature likely to be encountered has to be used as the basis for determining the current-carrying capacity. An alternative approach which may yield a more economic solution is to increase the cross-sectional area of the conductor only along that part of the route where the higher ambient temperature applies. *Regulations 523-1* and *523-2* give the basic requirements as regards suitability of a conductor for the ambient temperature likely to be met.

The remaining factor influencing the current-carrying capacity of a cable is the method of installation used and Table 9A of Appendix 9 of the Wiring Regulations lists a number of typical installation methods. Examination of the tables of current-carrying capacity in that Appendix shows immediately that the choice of installation method has a marked effect on current-carrying capacity. For example, from Table 9D2 it is seen that the current-carrying capacity for a twin-core or multicore cable installed in an enclosure such as conduit or trunking is in the order of 85% of that for the same cable mounted directly on a wall or embedded direct in ordinary plaster, the percentage difference varying with cable size. Sometimes one method of installation is used for part of a cable route and another method for the remainder, for example, part underground and part in air. In this case the current-carrying capacity which must be assumed is that corresponding to the more onerous method of installation, i.e. that method having the lower tabulated value of current-carrying capacity.

A significant reduction in current-carrying capacity also occurs if the cable of one circuit is installed in close proximity to, or bunched with, that of another circuit or of other circuits as indicated in Table 9B of Appendix 9. For instance, running the cables of two circuits in the same conduit reduces the current-carrying capacity of each circuit cable to 80% of that which would apply if installed in separate conduits. A particular example where it is sometimes forgotten that the grouping factor may have to be applied is that of cables in parallel. *Regulation*

*522-3* prescribes certain requirements intended, as indicated in that regulation, to see that the cables carry substantially the same fraction of the total current. Where cables connected in parallel are bunched together or run in the same enclosure, as they frequently are, the appropriate correction factor for grouping has to be applied.

It sometimes happens that along a cable route the number of other cables with which it is bunched (or with which it shares the same enclosure) varies and in determining the current-carrying capacity of the cable the grouping factor corresponding to the most arduous case, i.e. to the greatest number of cables bunched together anywhere along the route has to be used. Here again, an alternative approach is to increase the conductor cross-sectional area only at that part of its route where such bunching or grouping with other cables occurs.

Thus, in designing an installation it is sometimes necessary to establish whether it is commercially more attractive to instal the cables of a number of circuits in the same enclosure or subdivide them between two or more enclosures, the former method being the more expensive in terms of cable costs while the latter is more expensive in enclosure costs.

Sometimes when carrying out an extension or alteration to an installation an existing conduit or trunking is used to accommodate the additional cables needed but in such a case it is important to remember that the inclusion of those additional cables leads to a reduction of the current-carrying capacity of the cables already installed and a check should be carried out to verify that the reduction is not such that the operating temperatures become greater than those permitted by the Wiring Regulations.

It should be pointed out that as a general rule the grouping factors apply only if the conductors are not more than twice their overall diameters apart. The fact that any grouping which occurs is over a very short length does not obviate the need to apply the appropriate grouping factor and, for example, it may be necessary to use a number of separate entries to an enclosure in order to keep the cables concerned adequately separated so that the grouping factor need not be applied.

In practice, there may well be an admixture of cable sizes in a common trunking, cable ducting or conduit, and some may have increased cross-sectional areas in order to meet the voltage drop limitation prescribed in *Regulation 522-8* and are therefore not operating at the limiting temperature. Or, to quote another common case, lightly loaded control cables may be in the same enclosure as power cables and the imposition of the grouping factor corresponding to the number of circuits in the enclosure could lead to what may appear to be unnecessarily large sizes for the control cables. Nevertheless, unless it can be shown that the design method adopted does not lead to excessive operating temperatures, it may be better to re-examine the circuit arrangements in order to effect a subdivision of the circuits concerned in such a way that the appropriate correction factors can be chosen with confidence. In the case of control cables it is probably sufficient to apply only the temperature correction factor and ignore the correction factor for grouping.

The method of applying the various correction factors is not that previously used in the Wiring Regulations but is based on what is believed to be a more logical approach. It is felt that no matter what sequence of design steps is used, the designer will probably establish the design current of the circuit first and then determine the minimum conductor cross-sectional area to carry that current rather than start by choosing a particular conductor size and then determine the current it is capable of carrying.

In the Fourteenth Edition, the correction factors were applied as multipliers to the tabulated value of current-carrying capacity for a particular size of cable in order to obtain what could be termed the effective current-carrying capacity which had to be greater than the rated current of the overcurrent protective device it was intended to use and this in turn had to be greater than the design current of the circuit being designed. In other words, the cable size was picked first.

In the Fifteenth Edition, in following the more logical approach, the correction factors are applied *as divisors* to the rated current of the overload protective device, or if there is no such device, to the design current of the circuit. The minimum conductor cross-sectional area that can be used is that having a tabulated current-carrying capacity not less than the 'corrected' rated current of the overload protective device (or the 'corrected' circuit design current).

Thus, where overload protection is not required, the conductor used has to have a tabulated current-carrying capacity not less than

$$I_B \times \frac{1}{C_1} \times \frac{1}{C_2} \text{ A}$$

where

$C_1$ is the correction factor for grouping

$C_2$ is the correction factor for ambient temperature.

Where overload protection is to be provided by other than a semi-enclosed fuse to BS 3036 the tabulated current-carrying capacity has to be not less than

$$I_n \times \frac{1}{C_1} \times \frac{1}{C_2} \text{ A}$$

where $I_n$ is the rated current of the device.

If the overload protective device is a semi-enclosed fuse to BS 3036 a further correction factor must be introduced because of the relatively high fusing factor, as explained in the comments on *Regulation 433-2*, although this factor is not applied if it is intended to use mineral-insulated cables.

With that exception, when a BS 3036 semi-enclosed fuse is used the tabulated current-carrying capacity of the cable used has to be not less than

$$I_n \times \frac{1}{C_1} \times \frac{1}{C_2} \times \frac{1}{0{\cdot}725}$$

To take a simple numerical example, let it be assumed that a particular single-phase circuit, the design current of which has been calculated to be 22A, is to be wired in single-core p.v.c.-insulated cables to BS 6004 and these are to be installed in trunking with the cables of four other single-phase circuits. For the purpose of this example the design currents of the other circuits are not required to be known. It is further assumed that the designer has sufficient information to be able to assess that the ambient temperature will be 50°C.

If the circuit is protected against overload by a BS 1361 HBC fuse, the cross-sectional area of both the phase and neutral conductors are readily obtained as follows.

The nearest standard nominal current rating $I_n$ for the fuse, greater than $I_B$, must be selected; in this case $I_n$ is 30 A.

From Table 9B of Appendix 9,

$$C_1 = 0{\cdot}59 \text{ (because the total number of loaded conductors is } 5 \times 2 = 10)$$

From the ambient temperature correction factors to Table 9D1,

$$C_2 = 0{\cdot}71$$

so that the cross-sectional area of the phase and neutral conductors of the circuit must be equal to or greater than that which has a tabulated current-carrying capacity given by

$$30 \times \frac{1}{0{\cdot}59} \times \frac{1}{0{\cdot}71} \text{ A} = 71{\cdot}6 \text{ A}$$

Inspection of the relevant column of Table 9D1 immediately shows that a 10 mm² conductor has a current-carrying capacity of only 55 A and is therefore inadequate. The circuit has to be wired in cable having a conductor cross-sectional area of at least 16 mm².

If the circuit concerned is required to be protected against overload and it is intended to use a semi-enclosed fuse to BS 3036 for this purpose the cross-sectional area of the phase and neutral conductors must then be equal to or greater than that which has a tabulated current-carrying capacity given by

$$30 \times \frac{1}{0{\cdot}59} \times \frac{1}{0{\cdot}88} \times \frac{1}{0{\cdot}725} \text{ A} = 79{\cdot}7 \text{ A}$$

(0·88 is the correction factor for ambient temperature ($C_2$) for 50°C when a semi-enclosed fuse is used – see item 4 of the preamble in Appendix 9).

Inspection again of the relevant column of Table 9D1 now shows that when a semi-enclosed fuse is used the cross-sectional area of the phase and neutral conductors has to be increased to 25 mm².

## 9.2 Cables in thermal insulation

The increasing use of thermal insulation in buildings has led to the inclusion of

*Regulation 522-6*. Examination of the tables of current-carrying capacity in Appendix 9 of the Wiring Regulations shows that cables embedded in ordinary structural materials, such as brick or plaster, have the same current-carrying capacities as when the cables are 'open' and clipped direct to a surface, but this is only true if the structural materials are not of a type specifically designed and intended to provide thermal insulation. However, if a cable is either partly or wholly surrounded by thermal insulation or thermally insulating plaster the current-carrying capacity is lower compared with the 'open and clipped direct' value.

Where a circuit is lightly loaded, as may occur in final circuits for fixed lighting in dwellings, this reduction of current-carrying capacity due to thermal insulation may not be significant. For other types of final circuit it must be assumed that the actual loading may closely approach the current rating of the overcurrent protective device for the circuit and it is therefore essential to increase the conductor size to accommodate the reduction in current-carrying capacities as indicated wherever the installation of cables in thermal insulating materials cannot be avoided.

The precise correction factor to be used depends on the type and thickness of the thermal insulation and is not a constant. The factor also varies with the cable size and while further guidance concerning these correction factors is under consideration for incorporation in the Wiring Regulations, the approximate figures given in the Note to *Regulation 522-6* may be taken as a reasonable guide for types and thickness of thermal insulation now in common use and as averages over the range of cable size.

Thus where thermal insulation is used, either steps should be taken to avoid cable runs being installed in that insulation or, if this is not possible, to use a cable having a higher than usual current-carrying capacity so as to allow for the appropriate correction factor. For cables partly surrounded by thermal insulation this means the current-carrying capacity has to be 1·33 times the design current of the circuit or, if wholly surrounded, twice that design current. For the standard final circuits described in Appendix 5 of the Wiring Regulations this will generally mean an increase by one or two sizes, respectively, in conductor cross-sectional area.

With the present encouragement of the use of thermal insulation in buildings, both as loft insulation and in cavity walls, the presence of thermal insulation is normally to be expected and should be allowed for in the design of new or rewired installations unless, as already stated, deliberate steps are taken to avoid cable runs in the thermal insulation.

Fortunately, the most popular material (in the United Kingdom) used for cavity wall filling has been urea-formaldehyde foam and there is no evidence that this material has any adverse action with p.v.c., but expanded polystyrene granules used for loft insulation can cause migration of the plasticiser from the p.v.c. insulation of cables thereby reducing the flexibility of the cables. This loss of flexibility may of course, lead to a dangerous situation developing. While on this subject of compatibility, this migration of plasticiser from p.v.c. can lead to damage to the item with which the p.v.c. is in contact, the phenomenon being known as 'marring'. That damage can be slight, such as the spoiling of the surface appearance of the item or, more seriously, the damage can lead to considerable softening or cracking of the item concerned. Natural rubber grommets and polystyrene cable clips are two examples of items which may be softened by contact with p.v.c. whereas nylon, polypropylene and the majority of thermosetting plastics are considered to be satisfactory in contact with p.v.c.

## 9.3 Minimum ambient temperature

*Regulation 523-4* indicates that account need not be taken of the minimum ambient temperature in determining the normal operational conditions of conductors and cables but it also points out that precautions should be taken to avoid risk of mechanical damage to cables susceptible to low temperatures.

Cables insulated and/or sheathed with general purpose p.v.c. should not be installed in refrigerated spaces or other situations where the temperature is consistently below 0°C and if it is intended to install such cables, or cables which are paper-insulated or have bituminous compounded beddings and/or servings, during a period of low temperature, it is advisable to do so only when the ambient temperature is greater than 0°C and when the cable temperature is also greater than 0°C and has been so for the previous 24 hours.

## 9.4 Solar radiation

*Regulation 523-35* requires that cables and wiring systems subject to direct sunlight shall be of a type resistant to damage by ultraviolet light. For cables the best protection is given by black sheathing compounds.

In general, where weather resistance is a required feature of an installation cables sheathed with p.v.c. to BS 6746 or synthetic rubber to BS 6899 give adequate performance but natural rubber sheathing to BS 6899 normally has poor resistance to weather. Where polythene is used in outdoor locations, a black weather-resistant grade is essential.

In this context of cables exposed to solar radiation it must be borne in mind that the current-carrying capacity of a cable not shielded from the sun is less than that which would be obtained if the cable was shielded. To reduce the effect of solar radiation, cables, where possible, should be shielded from the direct rays of the sun without restricting ventilation.

Where this action cannot be taken, then as a general approximation, in order to take account of direct solar radiation, 20°C should be added to the ambient temperature and the appropriate correction factor used when determining the current-carrying capacity of the cable.

## 9.5 Colour identification

*Regulations 524-4* and *524-5* deal with the colour identification for cores of flexible cables and flexible cords and the details given in Table 52B of the colour identification are those now recognised throughout Western Europe. The Wiring Regulations identify the function of the cores of flexible cables and flexible cords by means of colour and the standard combinations of colours are those specified in BS 6004, BS 6007 and BS 6500, and are as follows:

| | | |
|---|---|---|
| Twin-core cables | : | blue and brown |
| Three-core cables | : | green-and-yellow, blue and brown |
| Four-core cables | : | green-and-yellow, black, blue and brown |
| Five-core cables | : | green-and-yellow, black, blue, brown and black. |

Taken together, the Wiring Regulations and the British Standards referred to give effect to CENELEC Harmonization Document HD 308.

The colour combination green-and-yellow, recognised in *Regulation 524-1* and Table 52A for protective conductors in fixed wiring, is also internationally agreed for that purpose. The requirements given in the remainder of Table 52A are the same as those of the Fourteenth Edition.

## 9.6 Lift installations

*Regulation 525-12* refers to a lift installation and it is important that the relationship between such an installation and the Wiring Regulations is made clear. A lift installation is considered to be a complete item of factory-built equipment complying as a whole with BS 5655, which corresponds to the European Standard EN 81: Part 1. Thus the detailed requirements of the Wiring Regulations apply to a lift installation only as far as the main intake of the supply to the lift machine room and to any fixed lighting in the lift well or machine room: the remainder of the installation has to comply with BS 5655.

Access to the lift shaft is undesirable for any persons other than those responsible for the lift installation and this is one reason no cables other than those of the lift installation are permitted to be run in a lift shaft.

## 9.7 Joints and terminations

Section 527 comprises twelve regulations concerned with joints and terminations and with one exception these appeared in the Fourteenth Edition. The exception is *Regulation 527-6* concerning compression joints. Compression or 'crimp-type' connectors are acceptable but, as indicated by the regulation, only if the correct tool is used for making the termination or joint. If, instead of the correct tool ordinary pliers were used, this could lead to high resistance joints and, as a consequence, premature failure of the connection.

When considering Section 527, it is impossible not to refer to *Regulation 13-1* which gives the general rule (applicable to all the installation and not only joints and terminations) that good workmanship and proper materials shall be used. Of all the components of an installation, probably the most indicative of the quality of workmanship employed are the joints and terminations. This is true not only of the regulations in this particular section which concern cable joints and terminations but also in relation to the joints and connections associated with metallic enclosures for cables particularly where such an enclosure is being used as the circuit protective conductor (see *Regulations 543-15* and *543-19*).

*Regulation 527-1* requires that terminations and joints shall not impose undue mechanical strain on the fixings of the connection or cause harmful mechanical damage to the conductors. Accessories such as ceiling roses and flexible cord outlet plates incorporate cord-grip devices and these should be correctly used to meet that Regulation.

Moulded connector strips and blocks are commonly used to make connections between cables in joint boxes and the like and are acceptable provided that they are of suitable material. Such connectors should be chosen with care so that they are of a material which will withstand the maximum temperature likely to occur in the enclosure in which they are situated (see *Regulation 523-1*).

As regards metallic enclosures for cables, it is essential that care is taken to obtain a high degree of electrical continuity and adequate capability to carry earth fault currents, not only when the installation is new but also subsequently. Cleaning of mating surfaces, the use of the correct conduit fittings and the prevention of rusting are all aspects which depend on the craft of the persons erecting the installation.

BS 4533 'Electric luminaires (lighting fittings)' permits luminaires in which conductors having basic insulation only are brought out of the luminaire for means of connection to the fixed wiring of the electrical installation. Such luminaires should only be selected for, and installed in, situations where those conductors are not accessible to persons other than those working on the installation. Such situations might be, for example, individual recesses on building structures or canopies housing luminaires and not requiring access by other trades, and these situations should take account of the requirements of *Regulations 523-22, 527-9* and *527-5*.

## 9.8 Cable capacities of conduit and trunking

*Regulation 529-7* is the only one in the Fifteenth Edition concerning the cable capacities of conduit cable ducting and trunking, requiring that the number of cables drawn into, or laid in, such enclosures shall not be so as to cause damage either to the cables or to the enclosures during installation.

Appendix 12 of the Wiring Regulations gives only guidance on the numbers of cables which can be accommodated in enclosures without contravening *Regulation 529-7*. Adoption of the 'unit system' and the tables in Appendix 12 does not lead to mandatory limits for the cable capacites of enclosures but is intended to indicate the numbers of cables it can be reasonably expected to be accommodated in a particular size and run of enclosure. The tables associated with the unit system are limited to single core p.v.c.-insulated cables in conduit or in trunking.

It is mentioned elsewhere in this Commentary that when designing the wiring system, e.g. determining the number of cables and the size of conduit to accommodate them, due account should be taken of the possibility of some future alteration to the installation, because the addition of cables to an existing conduit or trunking run will adversely affect the current-carrying capacity of the cables previously installed. In any event, it may be more attractive economically to subdivide the circuits between two or more enclosures in the first place in order to take advantage of the higher grouping correction factors for current-carrying capacity.

The method employing the unit system is very simple and is adequately explained in Appendix 12 of the Wiring Regulations so that here it is only necessary to show two worked examples of the method in operation.

### 9.8.1 Example 1

It is intended to run 8 single-core p.v.c.-insulated 2·5mm$^2$ cables and 6 similar cables but of 4mm$^2$ cross-sectional area. The estimated length of conduit run is 4 metres and includes 2 bends. What is the minimum size of conduit that can be used?

As the run includes bends, the cable factors have to be obtained from Table 12C and it is found that the factor for 2·5mm$^2$ is 30 and that for 4mm$^2$ is 43.

The sum of the cable factors therefore is

$$(8 \times 30) + (6 \times 43) = 498$$

Examination of Table 12D shows that for a 4 m run with 2 bends, a 25mm conduit has a factor of 388, and a 32mm conduit has a factor of 692 and therefore it is the latter size which has to be used.

### 9.8.2 Example 2

It is intended to run 16 single-phase circuits in trunking, each circuit having a design current of 12A. The ambient temperature is 30°C so that there is no correction factor for this but from Table 9B of Appendix 9 it will be found that the grouping correction factor for 32 loaded conductors is 0·39.

The overcurrent protective devices it is intended to use are BS 88 fuses so that the tabulated current-carrying capacity of the cables has to be at least

$$12 \times \frac{1}{0{\cdot}39} \text{ A} = 30{\cdot}8\text{A}$$

From Table 9D1 (as the cables are single-core p.v.c.-insulated cables) the minimum size which can be used is 4mm$^2$.

Assuming that each circuit has a protective conductor the same size as the phase conductors, there is a total of 48 conductors.

From Table 12E, the cable factor for $4mm^2$ single-core cables is 15·2 so that the trunking factor has to be at least 729·6, and from Table 12F it is immediately established that the minimum size of trunking which can be used is 75mm x 25mm or 50mm x 37·5mm.

If the circuits had been equally divided into two sets of eight, each set in separate trunking the grouping factor would increase to 0·51 and the cable size would reduce to $2{\cdot}5mm^2$. A cost comparison could be made to check which would be the cheaper method but the second choice has a disadvantage in as much as the trunking would be so lightly filled that at some future date additional cables could well be introduced. If they were, the $2{\cdot}5mm^2$ cables originally installed would be likely to become overheated, as they are (from Table 9D1) almost on the limit thermally with only the original 8 circuits.

Chapter 10

# Switchgear and other equipment

Little need be written here about *Regulation 531-1* because the requirements embodied in items (i) and (ii) have already been considered in Chapters 6 and 4, respectively, of this Commentary. The note to that regulation has also been considered in the chapter dealing with protection against electric shock – Chapter 4.

## 10.1 Residual current devices

*Regulations 531-3* to *531-8* concern residual current devices (r.c.ds). In the most common form, such as occurs in household and similar installations, the residual current device is what has hitherto been known as the current-operated earth leakage circuit breaker or core-balance circuit breaker but, as already indicated in this Commentary the term 'residual current device' covers any device which relies upon detection of residual current for its operation. The term therefore includes, for instance, the residual or zero-sequence-connected three protective current transformers (or a separate core-balance current transformer) which directly, or indirectly through a residual current relay, energise the trip coil of a conventional circuit breaker.

The modern generic term for the current-operated earth leakage circuit breaker is 'residual current circuit breaker' which is shortened to 'r.c.c.b', and there are various types of this device according to

(*a*) method of operation (with or without an auxiliary source)
(*b*) type of installation (fixed or mobile)
(*c*) number of poles and current paths, and
(*d*) additional (integral) protection (i.e. incorporating overcurrent protection)
(*e*) operating time (e.g. including time delay).

The additional (integral) protection referred to in item (*d*) may be for short circuit currents only or for both overload and short circuit currents. Where this protection is provided, it must be ensured that the rated making and breaking capacity of the r.c.c.b. is adequate compared with the prospective short circuit current at the point at which the r.c.c.b. is to be installed in order to protect the r.c.c.b. itself and afford compliance with Chapter 43 and Section 533 of the Wiring Regulations.

Where the r.c.c.b. it is intended to use does not have integral short circuit protection, a separate short circuit protective device is needed to ensure that adequate protection is given to the r.c.c.b. from the effects of short circuit currents up to the value of its rated conditional short circuit current. If the short circuit protective device is a fuse the manufacturer of the r.c.c.b. will be able to indicate its suitability from a knowledge of the fuse let-through $I^2t$ characteristic and cut-off current. This separate short circuit protective device will, of course, be that which is used to protect the circuit itself, and this particular aspect is dealt with in *Regulation 531-8*.

An earlier regulation, *Regulation 531-5*, draws attention to the fact that due account has to be taken of the leakage currents in normal service of current-using equipment being fed from the circuit or circuits protected by a residual current device (whatever its type) when determining what its rated residual current should be. Obviously there will be many cases where the installation designer is not in a position to know the sum total of these leakage currents for the very simple reason that he does not know what the current-using equipment is going to be and, if a number of circuits are to be protected by one r.c.c.b, how many of those equipments would be energised at the same time. Nevertheless, from a knowledge of the type of installation, he should be able to gauge whether or not the requirements of *Regulaton 531-5* will be met.

In terms of the limitation of insulation resistance to not less than 1 megohm, as prescribed in *Regulation 613-6*, this represents a leakage current of approximately 0·25 mA. Likewise, the maximum permitted leakage current for a fixed appliance not covered by a British Standard is approximately 0·5 mA, based on the requirement in *Regulation 613-8*.

The British Standards covering the safety requirements of equipment generally include limiting values of leakage current both at operating temperature and cold, in the latter condition after a specified time in a humidity cabinet. For instance, BS 3456 Part 101 which covers the general requirements for safety of household and similar electrical appliances prescribes the following limits, applicable to the two conditions:

| | | |
|---|---|---|
| (1) | For portable Class I appliances | 0·75 mA |
| (2) | For stationary Class I motor-operated appliances (only at operating temperature) | 3·5 mA |
| (3) | For stationary Class I heating appliances with heating elements that are detachable or can be switched off separately | 0·75mA or 0·75mA per kW rated input for each element or group of elements whichever is the greater, with a maximum of 5mA for the |

| | | |
|---|---|---|
| | | appliance as a whole |
| (4) | For other stationary Class I heating appliances | 0·75mA or 0·75mA per kW rated input of the appliance, whichever is the greater, with a maximum of 5mA |
| (5) | For Class II appliances | 0·25mA |

For appliances incorporating both heating elements and motors, the total leakage current is to be within the limit specified for heating appliances or for motor operated appliances whichever is the greater, but the two limits shall not be added. Part 3 of BS 3456 covers particular requirements for various types of household appliances and some of the standards in Part 3 vary the above limits.

For electrical commercial catering equipment BS 5784 Part 1, which gives the particular requirements for ranges, ovens and hot elements, in the main follows BS 3456 except that for items (3) and (4) the limit is 1mA (2mA cold) or 1mA (2mA cold) with a maximum of 10mA if the equipment is intended to be connected to the supply with a plug and socket or with no maximum if it is to be permanently connected to fixed wiring.

Section 533 prescribes requirements for overcurrent protective devices and as the reason for these requirements is self-evident it is not considered necessary to comment on them here but with one exception, namely, *Regulation 533-6* concerning discrimination.

## 10.2 Discrimination

*Regulation 553-6* retains the existing requirement in the Wiring Regulations, requiring that the characteristics and settings of overcurrent protective devices are such that proper discrimination in their operation is obtained. Discrimination can be defined as the ability of an overcurrent protective device to interrupt the supply to a circuit in which a fault has developed without affecting healthy circuits in the same system.

Probably the simplest practical example which can be chosen to illustrate discrimination is that of a ring circuit protected by a 30A (or 32A) fuse or miniature circuit breaker supplying a number of BS 1363 socket outlets. Should a fault develop in an item of current-using equipment or in its flexible supply cord it would be expected that the fuse in the plug would operate and not the circuit protective device, leaving the supply to the other socket outlets unimpaired, i.e. that there would be discrimination. It will be noticed that 'fault' is the word used because discrimination is necessary under both short circuit and earth fault conditions. While discrimination is a very interesting and important aspect of installation design, it is possible here in this Commentary to give only a brief outline of it.

Consider the circuit layout shown in Fig. 31*a*.

Here it is assumed that all the protective devices are HBC fuses; at the origin of the installation there is one of 800A rating and at the main distribution board there are three outgoing circuits each protected by a 400A fuse. For there to be discrimination, should a fault occur in any of these outgoing circuits, the fuse protecting that circuit (called the 'minor' fuse) must operate before the 800A fuse (the 'major' fuse).

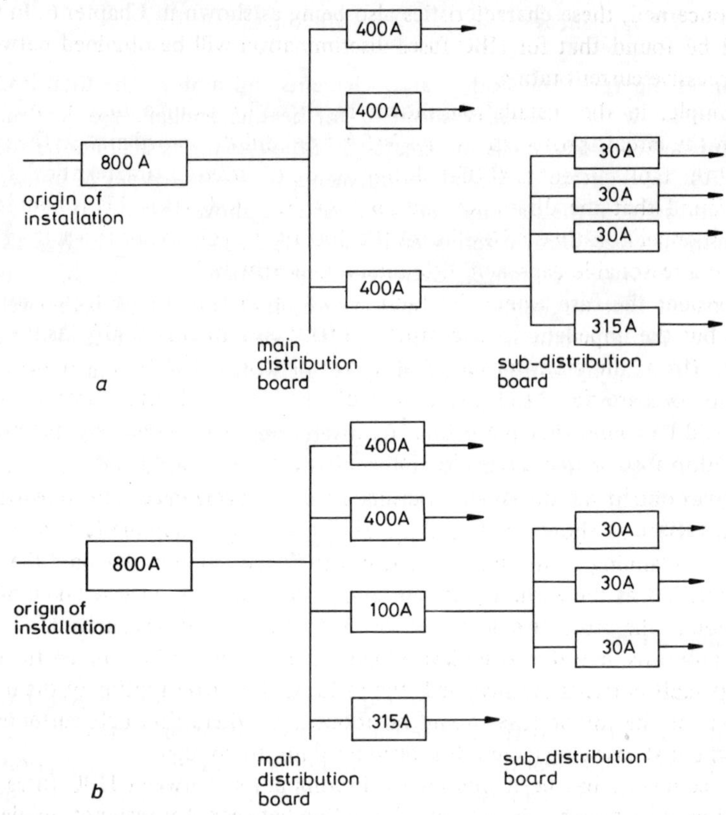


**Fig. 31** *Illustrating change of installation lay-out in order to achieve discrimination*

One of the outgoing circuits feeds a sub-distribution board and from this there is one final circuit protected by a 315A fuse and a number of final circuits each protected by 30A fuse. The 400A fuse at the sub-distribution board is now the major fuse and should not operate if a fault occurs in any of the final circuits.

In order to determine whether discrimination has been achieved one must first determine the prospective fault current in order to establish whether it is low, giving rise to pre-arcing times longer than 0·02s (assuming the system frequency is 50 Hz) or high, producing pre-arcing times shorter than 0·02 s. This differentiation is necessary because the answer will determine whether one can compare the time/current characteristics for the fuses concerned or if it is necessary to compare their $I^2t$ characteristics, (these latter characteristics have already been mentioned in Chapter 6 of this Commentary).

The key requirement for there to be discrimination between HBC fuses is that the total $I^2t$ of the minor fuse must not exceed the pre-arcing $I^2t$ of the major fuse. When the prospective fault current is low the difference between the pre-arcing $I^2t$ and total $I^2t$ for a particular fuse can be ignored and one can determine whether or not discrimination exists by comparing the normal time/current characteristics of the fuses concerned, these characteristics also being as shown in Chapter 6. In this case it will be found that for HBC fuses discrimination will be obtained between fuses of successive current rating.

For example, in the installation shown in Fig. 31*a*, assume that a fault has occurred in the circuit protected by the 315A fuse and the impedance to the fault is such that the fault current is 4000A. From the time/current characteristic for this fuse it is found that the disconnection time is 0·35 s and when this is compared with the disconnection time of 0·85 s for the 400A fuse, (the major fuse), it can be claimed that a reasonable degree of discrimination is achieved.

Now consider the case where the fault is still in the circuit protected by the 315A fuse but the impedance is such that the fault current is 10 000A. As the pre-arcing time (from the time/current characteristic) is less that 0·02 s it becomes necessary to compare the $I^2t$ characteristics of the 315A and 400A fuses. If this is done it would be found that the total $I^2t$ of the 315A fuse exceeds the pre-arcing $I^2t$ of the 400A fuse so that discrimination is not achieved.

In order to obtain the necessary discrimination it will be necessary to rearrange the system layout as shown in Fig. 31*b*. The very simple example just described shows that in planning an installation it is not sufficient to consider only the load requirements, but at the earliest stages in design due account must be taken of the co-ordination of the protective devices if the best results are to be achieved.

In general terms, in order to achieve a high degree of discrimination, particularly where large fault currents are involved, the ratio of the current rating of the major fuse to that of the minor fuse should be about 2:1. Where the fault currents are low, as in the first case considered, this ratio need not be so high.

So far, comment has been limited to discrimination between HBC fuses and consideration must now be given to discrimination between, for instance, miniature circuit breakers.

Fig. 32 shows the time/current characteristics of two miniature circuit breakers, one having a rated current of $I_A$ A and the other a greater rated current of $I_B$ A. Both miniature circuit breakers are assumed to be Type 1 and therefore they will trip within 0·1 s at four times their respective rated currents and at greater currents will instantaneously trip in a total operating time of approximately 0·01 s.

Thus, if the fault current is $I_{F1}$ A and if, as shown in Fig. 32, this is less than the instantaneous tripping current of the higher rated miniature circuit breaker, there will be discrimination whereas if the fault current is $I_{F2}$ A both circuit breakers would trip instantaneously and no discrimination exists. Much the same situation exists if one of the protective devices is a moulded case circuit breaker as these have similar operating characteristics as miniature circuit breakers.

Finally, in these brief comments on discrimination it only remains to consider

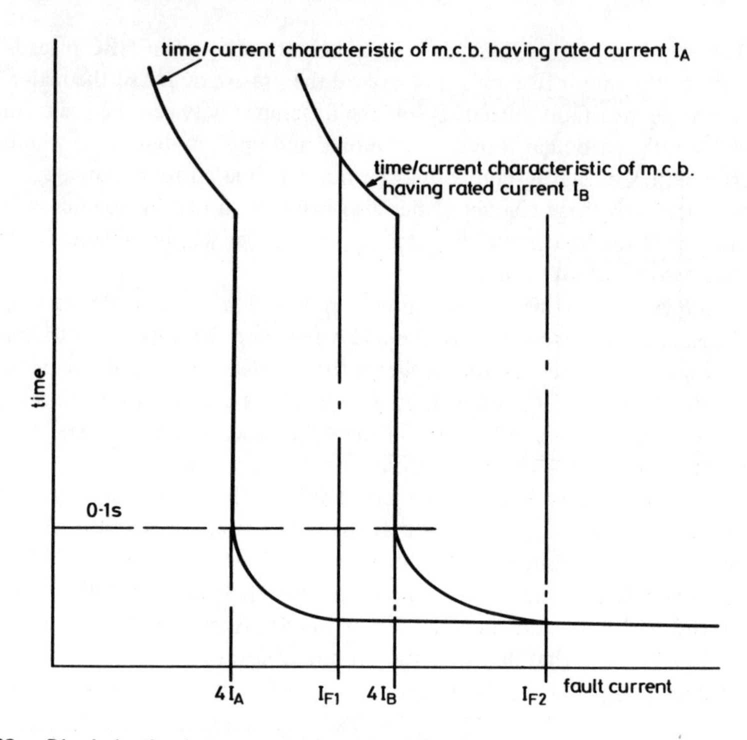


**Fig. 32** *Discrimination between miniature circuit breakers*

what happens if one of the protective devices is a fuse and the other is a miniature circuit breaker. The next Figure, Fig. 33, shows the time/current characteristic for an HBC fuse and that of a miniature circuit breaker having a lower rated current. The point of intersection between the two immediately indicates the maximum fault current for which the miniature circuit breaker will discriminate against the fuse.

It is of interest to indicate, as shown in Fig. 34, the degree of discrimination obtained when a circuit incorporating socket outlets to BS 1363 is protected by a miniature circuit breaker. In order to do this, it must be assumed that the fuse in the plug (or in a fused connection box) will be 13A and the miniature circuit breaker has a rated current between 15A and 32A.

*Regulation 533-6* refers specifically to discrimination between overcurrent protective devices and does not deal with discrimination between residual current devices of different rated residual operating currents when in series or between residual current devices and overcurrent protective devices. Bearing in mind the requirements prescribed in *Regulation 314-1* with regard to the avoidance of danger and inconvenience, due account should be taken of the need for discrimination when using residual current devices and in order to achieve this one method which can be adopted is to employ a residual current device which incorporates a time delay.

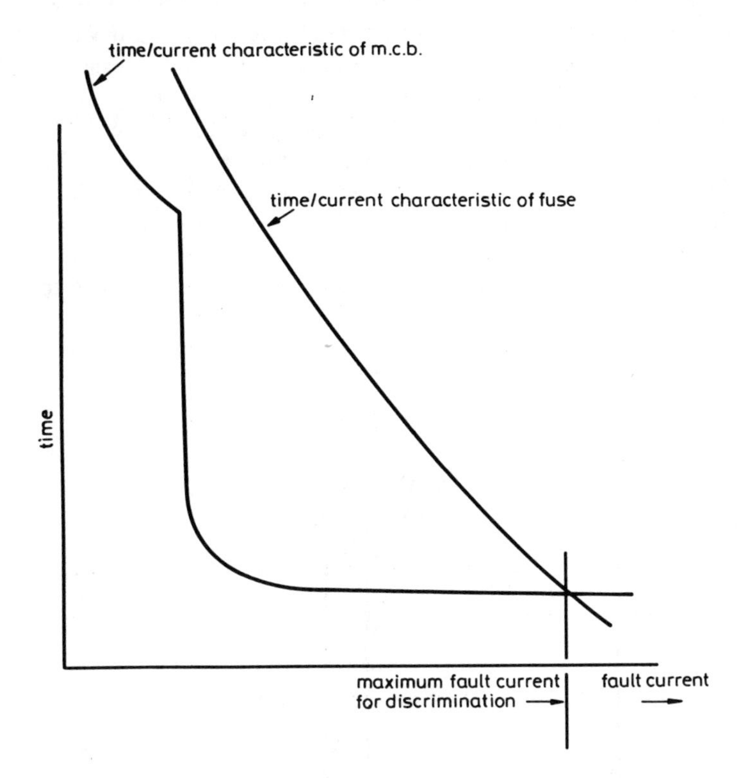


**Fig. 33** *Discrimination between miniature circuit breakers and fuses*

Section 537 details requirements for isolating and switching devices but in the main this section has been considered in Chapter 8 of this Commentary which dealt with isolation and the various forms of switching.

## 10.3 Rotating machines

*Regulation 552-1* concerns rotating machines and, mention has already been made elsewhere in this Commentary of the need, when determining the cross-sectional area of cables carrying the starting and accelerating currents of motors, to take account of the cumulative effect on the temperature rise of those cables caused by frequent stopping and starting. The Note to *Regulation 552-1* reminds the reader that it may be necessary to advise the supply undertaking of cases involving large motors in order to establish whether the starting and accelerating currents are within acceptable limits.

The supply undertaking has to be advised because large starting currents cause a so-called 'voltage dip' in the distribution network which may cause annoyance to other consumers fed by the same network, the degree of annoyance depending on the magnitude of the voltage dip and the frequency of its occurrence.

In order to assess whether in a particular case the annoyance to other consumers is within the acceptable limits, motors are divided into those subject to frequent starting and those subject to infrequent starting. 'Frequent starting' means starting at intervals of less than two hours.

For motors subject to infrequent starting, the voltage dip (phase to neutral) at

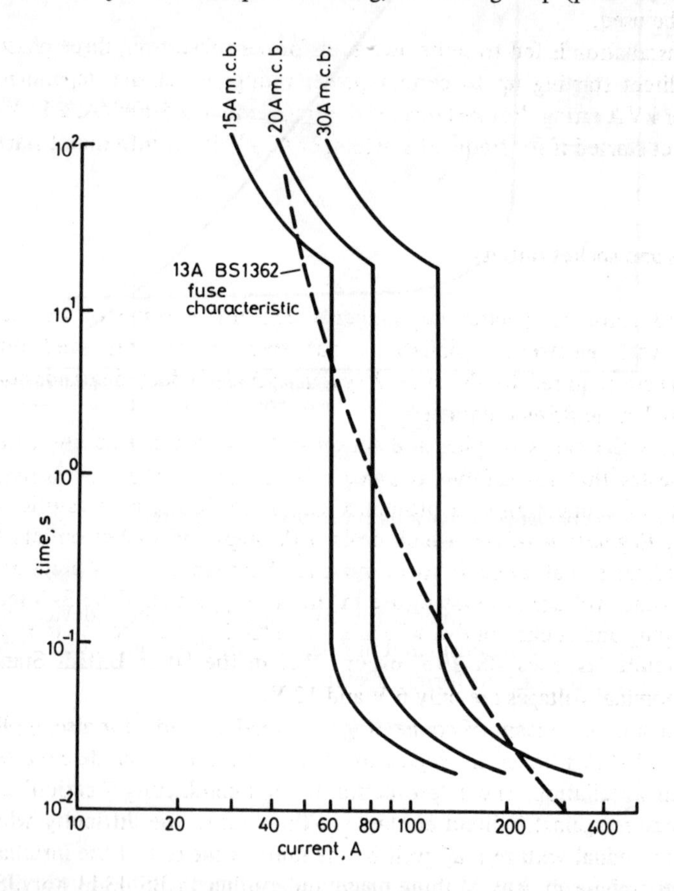


**Fig. 34** *Discrimination between miniature circuit breakers and the 13A BS 1362 fuse*

the point of common coupling with other consumers is generally required to be not in excess of 3%, and for motors subject to frequent starting the voltage dip limit is reduced to 1%. It is, of course, only the supply undertaking who can carry out this assessment as only they have the necessary information on relevant circuit impedances and can identify where the point of common coupling with other installations occurs.

In general where the supply is 240 V single-phase, motors up to 0·75 kW or, if 415 V three-phase, motors up to 4·5 kW, are acceptable without any qualification as regards the magnitude of the starting current or the method of starting, although if the motors are for lifts, hoists or similar applications the above power limits are halved. If the motors concerned have ratings in excess of the above limits it would be necessary to establish from the network and other relevant impedances the maximum starting current which can be tolerated without exceeding the 1% or 3% voltage dip limitation, as appropriate, and this will decide the method of starting which can be used.

If the installation is fed from its own transformer substation, three-phase motors can have direct starting up to certain power ratings which are dependent on the transformer kVA rating. For instance, if the transformer is 500kVA, a 15 kW motor can be direct started if for frequent starting, or 45 kW if for infrequent starting.

## 10.4 Plugs and socket outlets

Section 553 contains specific requirements concerning some of the accessories associated with electrical installations but these accessories (and others not mentioned) are required to also meet *Regulation 511-1* which demands compliance with the applicable British Standard.

As regards the types of plug and socket outlet which can be used, *Regulation 553-2* indicates that for safety extra-low voltage circuits, the general requirement for a means of connection of a protective conductor in these accessories does not apply, and *Regulation 411-8*, which deals with plugs and socket outlets for those circuits, indicates that there is to be no interchangeability with plugs and socket outlets of other voltage systems in use in the same premises. BS 4343 includes an industrial plug and socket outlet which are intended to be used in safety extra-low voltage circuits, as does BS 196, except that in the latter British Standard the intended nominal voltages are only 6 V and 12 V.

The omission of a means of connecting a protective conductor also applies to the 'special circuits' mentioned in *Regulation 553-3* and these include circuits in those parts of an installation where 'protection by non-conducting location' is the protective measure against indirect contact. In these cases, the difficulty which arises is that the nominal voltage may well be the same as the rest of the installation, e.g. 240 V single-phase or 415 V three-phase and neither in BS 4343 nor BS 196 are there suitable standard plugs and socket outlets. One is therefore forced to use a non-standard plug and socket outlet arrangement and the comment made against Note 3 to *Regulation 511-1* is pertinent.

*Regulation 553-6* deals with the plug and socket outlet arrangement for caravans and the distance of 20 m mentioned in this regulation is a compromise between the need to avoid excessively long flexible cables for connection of the supply to caravans and the facility given by *Regulation 471-44* to group the socket outlets on the site for up to six caravans. The quoted distance takes account of the varying positions of inlet connectors on caravans and the various positions in which the caravans may be parked on the site but whenever possible a shorter distance should be considered. The length of the flexible cable between the socket outlet and the caravan concerned should not be less than 10 m and not greater than 25 m.

It will be seen that *Regulation 553-7* does not admit BS 196 plugs and socket outlets for use on construction sites but requires that these accessories shall be to BS 4343. In this the Wiring Regulations follow BS 4363 which is the British Standard covering distribution units for use on such sites and CP 1017, the BS Code of Practice for electricity supplies on these sites.

The Wiring Regulations do not dimensionally specify the positioning of socket outlets (or for that matter, any other accessory) but *Regulation 553-8* indicates the need to mount a socket outlet in such a position as to avoid mechanical damage to that socket outlet and its associated plug and flexible cord. One must bear in mind that any size of flexible cord may be used up to that corresponding to the rated current of the plug, and it is therefore necessary to allow a distance such that the largest flexible cord likely to be used is not bent unduly sharply when the plug is inserted. For socket outlets mounted on a wall above a floor, allowance should be made for any floor coverings likely to be fitted subsequently and account should also be taken of mechanical damage likely to be caused if the socket outlet and plug are hit by floor cleaning equipment. In the absence of detailed knowledge of the circumstances likely to prevail in practice it is recommended that a mounting height of 150 mm is used although in some circumstances a greater height may be desirable from the aspect of convenience of use.

It should also be borne in mind that socket outlets, as all other equipment, are required to comply with *Regulation 512-6* and must therefore be of a type appropriate to the situation in which they are used and their mode of installation must take account of the conditions likely to be encountered. There is no regulation which forbids, for instance, the mounting of socket outlets in a floor but where this is done it must be ensured that the floor can be washed without risk of shock or detriment to the insulation.

The Fifteenth Edition, like its predecessors, does not specify the number of socket outlets to be installed in a particular location, but it does recognise that there is a safety aspect, the intent of *Regulation 553-9* being to ensure that the number of socket outlets is sufficient to avoid the use of unduly long flexible cords, adaptors and extension leads. The note to this regulation draws attention to the fact that flexible cords fitted to appliances and luminaires are usually 1·5 m to 2 m long, and this fact suggests that in determining the position of socket outlets in a room in a dwelling one should aim for a distance between socket outlets not exceeding 3 m. Regard should also be given to the positioning of socket outlets so that appliance flexible cords are not usually required to trail across doorways and the like.

Recommendations on the number of socket outlets to instal are made in 'Homes for today and tomorrow' (the 'Parker Morris' Report, 1961, HMSO 75-90), and in publications issued by the Department of Environment (1971), the Electrical Contractors' Association (1974), the National House Building Council (1974), the Institution of Heating and Ventilating Engineers (1974; now Chartered Institute of Institute of Building Services), and the Electrical Installation Industry Liaison Committee (1975).

The only statutory requirements are those specified, for housing, in Part Q of the Building Standards (Scotland) (Consolidation) Regulations 1970.

Chapter 11

# Inspection and testing

## 11.1 Chapter 61: Initial inspection and testing

The Fifteenth Edition of the Wiring Regulations and its Appendixes contain a number of simplified standard methods of compliance which are intended to be suitable for many common cases, and which are of particular assistance in designing household and similar installations. Where these simplified systems are used, verification of compliance is a simple matter of checking by inspection or test that the specified standard arrangements have been complied with.

However, as indicated at the beginning of this Commentary, the basic regulations for the Fifteenth Edition allow the designer greater freedom than hitherto by which he may, if he wishes, comply by calculation from simple formulae. One example of this is the use of the formula given in *Regulation 434-6* by which means the designer, after having calculated the prospective short circuit current applicable to a particular circuit, may determine that the overcurrent protective device operates sufficiently rapidly to limit the temperature of the circuit conductors to an acceptable value, such determination merely being inspection of the relevant time/current characteristic.

Where the designer relies upon calculation, he should be prepared to justify his calculations because these would be the major means of verification of compliance. For example, the checking of the designer's calculations, in some cases, may be the only method of establishing that the requirements of *Regulation 543-2* as regards the cross-sectional areas of protective conductors had been met.

*Regulation 611-1* includes the important requirement that the methods of test which are used shall themselves cause neither danger to persons nor damage to equipment even if the circuit being tested is defective. It is therefore important that the sequence of testing detailed in *Regulation 613-1* is followed, to ensure that the protective conductors in the installation are correctly connected and electrically sound before the installation is energised and before any other tests involving these conductors are made. The sequence is also intended to see that all equipment of the installation is correctly installed before the verification of polarity, the earth loop impedance tests and the correct operation of residual current devices.

### 11.1.1 Visual inspection

*Regulation 612-1* requires that a visual inspection has to be made of an installation, the first aim of that inspection being to confirm that all equipment complies with the relevant British Standards. Where equipment does not carry any mark or certification indicating compliance with a British Standard (or an equivalent foreign standard) reference should be made to Note 3 of *Regulation 511-1*. For equipment which is the subject of an ANTS Certificate, the number of that Certificate should be entered in the space provided in both the Inspection and Completion Certificates (see Appendix 16 of the Wiring Regulations). Where the equipment is not covered by a British Standard or by an ANTS Certificate, if there is any doubt as to its compliance with the Wiring Regulations the designer should be ready to produce evidence of the action he has taken to verify compliance.

Visual inspection of an installation is carried out also for the purpose of checking that equipment has been correctly selected and erected in accordance with the Wiring Regulations and Appendix 14 outlines the principal features which should be visually inspected.

Particular note should be taken of the requirements of *Regulation 542-20* with regard to the accessibility and disconnectibility of the earthing arrangements which may conveniently include the means of disconnecting the main bonding conductor, as required, in order to carry out the earth fault loop impedance test described in Item 5 of Appendix 15 of the Wiring Regulations.

Another aspect which must not be forgotten is that verification of compliance with the requirements prescribed for the measure 'Automatic disconnection of the supply' may be achieved by visual inspection of the protective device concerned to ascertain that it is of a standard type and of the correct current rating; for this measure it must, of course, be also checked that the measured impedances do not exceed the maximum permitted values given in the relevant table in *Regulation 413-5*, Appendix 8 or, where the alternative method described in Appendix 7 is used, that the impedance of the protective conductor does not exceed the maximum permitted value given in Appendix 7.

If the cross-sectional areas of protective conductors have been determined from the formulae in *Regulation 543-2* and Appendix 8, the visual inspection should include confirmation of the sizes of protective conductor used; similar confirmation is required if the cross-sectional areas of the live conductors have been determined from the formula in *Regulation 434-6*.

It should be noted that the Inspection Certificate now includes two items directly related to these design aspects concerning phase and protective conductors, the first of these items being the prospective short circuit current at the origin of the installation and the second being the earth fault loop impedance, also at that origin. While the terms of the Inspection Certificate are such that it need only be indicated that the values of these two supply aspects are 'satisfactory' it is suggested that the actual values be recorded on the certificate. Similarly, the actual earth fault loop impedance for each final circuit should also be recorded.

It will be appreciated that the checklist given in Appendix 14 is neither exhaustive or detailed and it is suggested that the installer, using the items in that Appendix as 'headings' could develop detailed lists of the factors needing attention when installing and inspecting equipment.

Another useful approach which could be adopted would be to prepare detailed lists of the factors for particular components of the installation, segregating those components, for example, into

(i) non-flexible cables
(ii) flexible cables and cords
(iii) cable enclosures
(iv) wiring accessories
(v) fixed equipment

Obviously item (iv) could be subdivided, for example, into

(a) general requirements for accessories
(b) socket outlets
(c) lighting switches
(d) ceiling roses
(e) cooker control units . . . . . . . . . . and so on.

Inspection lists prepared by either method are also useful to the designer and while the precise form of any such list is a matter of personal preference the following is offered as one example, being a list of requirements concerning metallic conduit.

(1) Complies with relevant British Standard, i.e. BS 31 or BS 5468 (*Regulation 521-9*)
(2) Completely erected before cables are drawn in (*Regulation 521-10*)
(3) Connected to main earthing terminal (*Regulations 413-8, 413-10* or *413-14*)
(4) Segregated from, or bonded to other fixed metalwork (*Regulation 525-10*)
(5) Properly supported and suitable as regards mechanical damage, either by type or by protection (*Regulation 529-2*)
(6) If in damp situation or exposed to weather, is corrosion resistant by type or by finish (*Regulation 523-8*)
(7) If not sealed system, is provided with drainage holes (*Regulation 523-14*)
(8) Used in association with cables for overhead wiring (*Regulation 523-26*)
(9) Used in association with flexible cords (*Regulation 523-29*)
(10) Inaccessible to livestock (*Regulation 523-34*)

(11) Colour identification - basic colour orange (*Regulation 524-2*)
(12) When accommodating Category 1 and Category 2 circuits together (*Regulations 525-3* and *525-7*)
(13) Not to accommodate Category 1 and Category 3 circuits together (*Regulation 525-4*)
(14) Not to accommodate electrical and other services together (*Regulation 525-11*)
(15) Freedom from burrs and avoidance of damage to cables (*Regulation 527-10*)
(16) Junction boxes in conduit system (*Regulation 527-11*)
(17) Connection with other wiring systems (*Regulation 527-12*)
(18) When passing through fire barriers (*Regulation 528-1*)
(19) Use of solid elbows and tees (*Regulation 529-4*)
(20) Radii of conduit bends (*Regulation 529-5*)
(21) Space factor (*Regulation 529-7*)
(22) When used as a protective conductor (*Regulations 543-9* and *543-10*)
(23) Exemption for conduit from *Regulation 526-1* as regards accessibility of joints (*Regulation 543-16*)
(24) Joints to be mechanically and electrically continuous (*Regulation 543-19*)
(25) *k* factor for conduit (*Regulation 543-2* Table 54B)
(26) Spacing of supports (Appendix 11, Item 4 and Table 11C)
(27) Non-sheathed cables, rubber- or p.v.c.-sheathed cables in conduit (Appendix 11, Items 1 and 8)
(28) Cables in conduit spanning building (Appendix 11 Item 17 and Table 11B)
(29) Protection against corrosion (Appendix 10).

The above list details only those regulations directly related to metallic conduit and it must also be established that the size of conduit used is such that the number of cables drawn into it is not so large that damage may be caused to those cables when being pulled in. Furthermore, the conductors of all phases and neutral (if any) of a particular circuit must be contained in the same conduit. The visual inspection should include the determination of the number of conductors in conduit (or any other type of cable enclosure) to establish that the correct grouping factors have been applied.

### 11.1.2 Testing

Section 613 deals with the testing of installations and as already stated the methods adopted must not cause damage to equipment or danger to persons. *Regulations 613-5* to *613-8* cover the insulation resistance tests which must be carried out and among this series *Regulation 613-7* draws particular attention to the presence of electronic devices in an installation and their susceptibility to damage from the test voltages prescribed in *Regulation 613-5*.

There is an increasing use of equipment having electronic components; items of equipment such as dimmer switches, touch switches, delay timers, power controllers as well as control circuits incorporated in current-using equipment, including luminaires. If possible, the insulation resistance tests should be carried out before such devices are connected but if this cannot be done the devices should be temporarily disconnected so that they are not subjected to the test voltage.

### 11.1.3 Continuity of ring final circuit conductors

*Regulation 613-2* concerns the continuity of ring final circuit conductors and Item (2) of Appendix 15 details two methods of test. It is accepted that neither of these tests is completely infallible as regards the verification of continuity, but they are an attempt to offer practical tests which can be readily and speedily carried out and at the same time give the resistance values for the ring circuit conductors which can then be used for the determination of the earth fault loop impedance $Z_s$ as demanded by *Regulation 613-15* or for the alternative method described in Appendix 7 of the Wiring Regulations.

### 11.1.4 Test equipment

*Regulations 613-10* and *613-12* specify tests which call for equipment not hitherto required by the Wiring Regulations, namely, Test Finger I to BS 3042 (see Appendix B of this Commentary) and a straight stiff steel wire 1mm in diameter (also fully specified in Appendix B). However, installations involving the need for these tests occur only in exceptional cases, and as indicated elsewhere in this Commentary, wherever possible the use of factory-built equipment incorporating its own means of protection against direct contact is preferred.

For *Regulation 613-10*, a high-voltage tester is also required, suitable for the applied high voltage test specified in the British Standard concerned. A high-voltage tester may also be required in order to prove compliance with

(*a*) *Regulation 411-3* concerning the safety source of a safety extra-low voltage circuit

(*b*) *Regulations 411-6* concerning the electrical separation of a safety extra-low voltage circuit from other circuits

(*c*) *Regulations 411-12* and *411-13* concerning functional extra-low voltage circuits

(*d*) *Regulation 413-22* concerning the insulating enclosure used in the protective measure 'Protection by use of Class II equipment or by equivalent insulation'

(*e*) *Regulation 413-36* concerning the source of supply to a circuit where the protective measure for protection against indirect contact is 'Protection by electrical separation'.

As regards item (*b*), it will be seen from Chapter 4 of this Commentary that the high-voltage tester must be capable of giving a test voltage of 4kV when testing equipment having a nominal voltage of 380V or less, or 6kV when that nominal voltage exceeds 380V.

### 11.1.5 Non-conducting location

*Regulation 613-13* concerns the testing of the floors and walls of a location where the protective measure against indirect contact is 'Protection by non-conducting location' to check that the resistances of the floors and walls to the main protective conductor of the installation (no protective conductors are permitted inside the location) do not exceed the limits specified in *Regulation 413-29*. It will be necessary to construct suitable test equipment which provides adequate contact between the electrode of that equipment and the points on the surfaces to be tested and to which the electrode is applied.

### 11.1.6 Polarity

*Regulation 613-14* requires testing for polarity and it should be noted that *tests* to determine polarity must be made. Visual inspection alone is not sufficient as this gives conductor identification only.

### 11.1.7 Earth fault loop impedance

The measurement of earth fault loop impedance is required where the protective measure against indirect contact involves automatic disconnection of the supply and, as indicated in Item 5 of Appendix 15 of the Wiring Regulations, there are two basic methods one can adopt to check whether the values of earth fault loop impedances actually obtained from the circuits of an installation are within the limits prescribed in the relevant regulations.

Attention is drawn to the tabulated limiting values of earth fault loop impedance given in Chapter 41 and Appendices 7 and 8 of the Wiring Regulations which are such that the instruments used to measure earth fault loop impedance (and for the method detailed in Appendix 7, the resistance of the protective conductor) should be sufficiently accurate.

The Fifteenth Edition differs from preceding editions because, as stated in Item 5 of Appendix 15, when measuring the earth fault loop impedance external to the installation $Z_E$, the main equipotential bonding conductors are temporarily disconnected, but when determining the total earth fault loop impedance $Z_s$ at any point inside the installation, these conductors are retained in place.

When measuring the earth fault loop impedance of a circuit or installation protected by a fault-voltage operated device, the conductors bonding extraneous conductive parts to the frame terminal of that device must be temporarily disconnected in order to ensure that the measured impedance relates to the fault current path through the associated earth electrode.

An equally important difference is that the Fifteenth Edition recognises a method of determining the earth fault loop impedance $Z_s$ without using an earth fault loop tester within the installation, and this is of particular significance where the installation includes residual current devices.

This method is limited to circuits where the conductors have a cross-sectional area of $35\text{mm}^2$ or less, and entails measuring first of all the earth fault loop impedance external to the installation. For each radial circuit a measurement is then made of the resistance of the phase conductor $R_1$ from the point of origin of the circuit to the most distant point of utilisation supplied by the circuit in series with the resistance of the circuit protective conductor $R_2$ from the latter point back to the origin.

The earth fault loop impedance $Z_s$ is then taken to be the arithmetic sum of $(Z_E + R_1 + R_2)\ \Omega$. As stated in Appendix C of this Commentary, this value of $Z_s$ is pessimistically high because $Z_E$ will not be wholly resistive as inferred by the arithmetic nature of the summation and, because the main equipotential bonding conductors are not connected during the test to measure $Z_E$, the measured value of that impedance will be high due to the loss of the earth fault paths afforded by water pipes, gas pipes and other extraneous conductive parts.

For non-spurred ring circuits, the method is slightly different because the resistance $R_{t1}$ of the complete length of the phase conductor in series with the resistance $R_{t2}$ of the complete length of the protective conductor is measured. The earth fault loop impedance is then taken as

$$Z_s = Z_E + \frac{(R_{t1} + R_{t2})}{4}\ \Omega$$

For more information on ring final circuits see Appendix D of this Commentary.

For a ring circuit having spurs, either one proceeds exactly as for the non-spurred case by adding to the above value for $Z_s$ the resistances of the phase and protective conductors for the longest spur or else the effective series/parallel resistance is measured between the extremity of each spur and the origin of the ring circuit.

This 'summation' method of determining the earth fault loop impedance is preferred when the circuit is protected by a residual current device, because if the earth fault loop impedance is measured directly, any such device would have to be temporarily removed from the circuit or its main circuit terminals short-circuited (i.e. bridged). The summation method has to be used where compliance with *Regulation 413-4* is obtained by limitation of the resistance of the circuit protective conductor as described in Appendix 7 of the Wiring Regulations. The earth fault loop impedance $Z_s$ is still required to be known and the one measurement of the circuit phase and protective conductors in series is still sufficient where these are of the same material and follow the same route.

The ratio of the resistance of the protective conductor $R_2$ to that of the associated phase conductor $R_1$ is equal to the inverse ratio of the respective conductor cross-sectional areas and is denoted by '$m$', i.e.

$$\frac{R_2}{R_1} = \frac{A_1}{A_2} = m$$

Appendix D of this Commentary includes a table giving the values of $m$ for the flat cables complying with BS 6004.

Having measured $R_1$ and $R_2$ in series, $R_2$ can then be calculated from

$$R_2 = (R_1 + R_2) \frac{(m)}{m+1} \, \Omega$$

and this must not be greater than the tabulated value given in Appendix 7 of the Wiring Regulations appropriate to the type and current rating of the overcurrent protective device concerned.

The measurement of the resistances of the phase and neutral conductors in series also has its uses in those cases where it is necessary to check that the circuit concerned complies with the thermal requirements related to short circuit conditions, as prescribed in *Regulation 434-6*.

For example, assume that the installation is single-phase and that the nominal voltage is $U_n$ V. Also assume that the neutral conductors of the final circuits concerned have the same cross-sectional areas and are of the same material as the phase conductors, as will normally be the case. The prospective short circuit current at the origin of the installation has been ascertained as required by *Regulation 313-1* and if this is denoted by $I_{sc}$, the impedance of the phase-neutral loop external to the installation $Z_c$ is given by

$$Z_c = \frac{U_n}{I_{sc}} \Omega$$

As indicated in Chapter 6 of this Commentary, in order to check compliance with *Regulation 434-6* (where it has not been possible to invoke *Regulation 434-5*), it is necessary to determine the minimum short circuit current which can occur so that for a particular circuit it is necessary to determine the phase-neutral loop impedance at the point of utilisation furthest from the origin of the circuit.

Thus, having measured $R_1$ and $R_2$ in series, as already explained, $R_1$, the resistance of the phase conductor, is readily obtained from

$$R_1 = \frac{R_1 + R_2}{1 + m} \, \Omega$$

So that the maximum phase-neutral loop impedance is given by

$$\frac{U_n}{I_{sc}} + \frac{2(R_1 + R_2)}{1 + m} \, \Omega$$

Having determined this impedance, the minimum short circuit current is calculated and inspection of the time/current characteristic of the overcurrent protective device will indicate whether the disconnection time is sufficiently rapid to give compliance with *Regulation 434-6*. It is also necessary to determine the maximum short circuit current in order to ensure that the protective devices being used have an adequate breaking capacity rating as required by Chapter 43 of the Wiring

Regulations and, as illustrated by Fig. 11 of this Commentary, with some types of circuit breaker if the short circuit current is too high the thermal requirements of *Regulation 434-6* will not be met.

### 11.1.8 Residual current devices

*Regulation 613-16* requires the test described in Item 6 of Appendix 15 of the Wiring Regulations to be carried out on any residual current device incorporated in the installation. The test is that described in the Fourteenth Edition and while it is a proven method it does not give any indication of the time of operation. It is understood that test equipment is available which is capable of verifying that operating times are within the values prescribed in the Wiring Regulations in *Regulation 413-4*.

## 11.2 Chapter 62: Alterations to installations

Chapter 62 of the Wiring Regulations comprises only two regulations and it is the second of these (*Regulation 622-1*) which is commented on here.

Where an alteration to an installation is ordered, it is the responsibility of the person carrying out that work to see that the alteration complies with the Wiring Regulations. Where compliance is prevented by a defect in part of the existing installation which is related to, but forms no part of, the alteration work, the alteration should not be connected and the person ordering the work should be informed of the defect, otherwise *Regulation 13-19* would be contravened. For this reason, before undertaking an alteration, a preliminary inspection of the existing installation should always be made.

It must also be verified that the cables, protective devices and earthing arrangements of the existing installation would be adequate to maintain compliance with the Wiring Regulations after the alterations or additions have been completed.

Chapter 12

# The smaller installation and the Fifteenth Edition

For the purposes of this final chapter of the Commentary, the 'smaller installation' is defined as one where the maximum demand does not exceed 100A, if single-phase, or 100A per phase if three-phase, and is fed directly from the public supply low voltage network. The term therefore embraces the great majority of electrical installations and includes office, shop and industrial installations as well as those for dwellings.

For very many electrical contractors the term also represents the *only* type of installation of which they have experience and the aim of this chapter is to assist them in the implementation of the requirements of the Fifteenth Edition in such installations.

What this chapter does *not* set out to do is to provide sufficient information for the contractor to design an installation without him having to read and understand the Wiring Regulations, nor is it intended to be a substitute for the other chapters of this Commentary although it will be found that there is some reiteration here of certain points made in those chapters.

## 12.1 Types of cables, their identification and methods of erection

Before commenting on the aspects of the Fifteenth Edition which differ from preceding editions, it can be stated immediately that as regards the types of cables that can be used, their identification and the permissible methods of erection prescribed in the new edition there is nothing which prohibits existing practices. Chapter 52 of the Fifteenth Edition is essentially the same as Section B of the Fourteenth Edition but it is believed that the requirements are now more logically presented, being placed under section and other headings which indicate the reason for any particular requirement.

Similarly it will be found that, although the method of determining the cross-sectional area of a conductor to use for a given maximum demand and type of overcurrent protection is not the same as in previous editions, in a great many cases the Fifteenth Edition requirements will not lead to increased conductor sizes.

## 12.2 Meter tails

For household and the smaller office and shop installations, the normal service provided by a supply undertaking will be 240V single-phase to a meter board and the undertaking's equipment will include an all-insulated service cut-out incorporating a BS 1361 fuse of 60A, or now more commonly, 100A, current rating.

Consumer's equipment must not be fixed to the supply undertaking's meter board. For new properties it is becoming increasingly popular for supply undertakings to use external meter cabinets fixed in an outside wall of the buildings and it is understood that in most cases the supply undertakings will not allow consumer's equipment to be fitted in these cabinets.

Although the purpose of the fuse in the supply undertaking's cut-out is to protect the undertaking's own supply cables and equipment as required by the Electricity Supply Regulations, the Note to *Regulation 13-7* recognises the probable acceptance by the undertakings that the fuse may be considered to give protection against short circuit currents in the consumer's cables (i.e. the meter tails) between the meter and the consumer's control unit provided that these cables are reasonably short. It is generally accepted that 'reasonably short' in this context means not more than three metres.

A supply undertaking may have specific requirements as regards the type of meter tails to be installed and it is the responsibility of the installation designer to determine what these requirements are, but in any event the requirements of the Fifteenth Edition must be met.

Meter tails are normally single-core p.v.c.-insulated and sheathed cables to BS 6004 except that if they are considered likely to be subjected to mechanical damage they must be adequately protected throughout against such damage (*Regulation 523-19*) by conduit or trunking, and then do not require to be sheathed.

The cross-sectional area of the meter tails has to be such that the current-carrying capacity is not less than the maximum sustained current passing through them (*Regulation 522-1*), i.e. the maximum demand of the installation. When the supply undertaking's fuse is 60A it is therefore usual to instal 16mm$^2$ meter tails or, if 100A, to instal 25mm$^2$ meter tails and these sizes comply with Table 9D1.

If however, the meter tails are installed in close proximity with other cables, it will be necessary to apply the appropriate grouping factor (Appendix 9) and increase the cross-sectional area accordingly. Here it is pointed out that 'whole current' meters do not accept meter tails greater than 35mm$^2$, and if larger tails are required this would mean the use of current transformers by the supply undertaking. This is another reason why close liaison at the earliest stages of a project between the supply undertaking and installation designer is extremely important.

The sizes of cables referred to are for copper conductors and generally it will be found that supply undertakings will not allow the use of cables with aluminium conductors for meter tails.

The fuse ratings quoted earlier will usually be adequate for the installations being considered in this chapter but it is still the responsibility of the installation designer to assure himself that the supply being offered by the undertaking is in fact adequate by assessing the maximum demand (*Regulations 311-1* and *311-2*).

## 12.3 Assessment of the characteristics of the supply

Before he can proceed with the detailed design of the installation the designer has also to carry out the other assessments prescribed in Part 3 of the Wiring Regulations but here it can be assumed for the purposes of this chapter that the supply is single-phase 240V phase to neutral or 415V three-phase line to line, the frequency being 50Hz. Thus, as required by *Regulations 312-2* and *313-1*, he has also to ascertain from the supply undertaking

(*a*) the type of earthing arrangement used for the supply
(*b*) the earth fault loop impedance external to the installation, and
(*c*) the prospective short circuit current at the origin of the installation.

## 12.4 Type of earthing arrangement

As regards the first of these items, it is expected that on most, if not all, new supplies, a PME earthing terminal will be offered by the supply undertaking but this may be used by the consumer only if the installation complies with the requirements of the PME Approval 1974. Note 2 to *Regulation 413-2* indicates that compliance with that regulation also generally satisfies the PME Approval.

The other earthing arrangement encountered is the provision of an earthing terminal from the lead sheath of the supply undertaking's cable.

## 12.5 Earthing conductor

The first problem encountered by the installation designer when either of the above earthing arrangements is available is the sizing of the earthing conductor, this being the conductor connecting the means of earthing to the earth bar in the consumer's distribution unit.

The designer has two possibilities. He can implement Table 54F which means that if the meter tails are $16mm^2$, $25mm^2$ or $35mm^2$ the earthing conductor has to be at least $16mm^2$. If the meter tails are greater than $35mm^2$, the earthing conductor has to have a cross-sectional area at least half that of the meter tails. If the meter tails are less than $16mm^2$, the earthing conductor has to be the same size as them.

The other approach he can adopt is to use the equation given in *Regulation 543-2* with the appropriate factor from Table 54B, and in order to do this he has to determine the earth fault loop impedance at the incoming terminals of the consumers' unit, hence the need to obtain the value of that part of the earth fault loop impedance which is external to the installation.

This information is required in order to design the circuit protective conductors throughout the installation but for the earthing conductor, unless there is some difficulty in accommodating the cable size in the terminal or bar provided in the consumer unit, it is suggested that the Table 54F approach should be used.

The value of the earth fault loop impedance external to the installation, denoted by $Z_E$ Ω, has to be obtained from the supply undertaking concerned, but it must be appreciated that they are under no statutory duty to guarantee this and, in practice, will not be able to do so because they may modify their supply network, as appropriate, to meet the changing loads of consumers.

The supply undertaking may indicate a range of values of $Z_E$ likely to be obtained, in which case the maximum value in that range should be assumed.

The following values of $Z_E$ are believed to be the maxima likely to be encountered except in extreme situations. If the supply is already available $Z_E$, of course, can be measured, otherwise the following figures can be used for design purposes after having checked with the supply undertaking concerned that these values are realistic for the installation being designed.

When the supply is PME, it can be assumed in most cases that $Z_E$ will not exceed 0·35 Ω, this value being based on a 315 kVA supply undertaking's distribution transformer, a 500 m length of $95mm^2$ aluminium conductor mains cable and a 10 m length of $25mm^2$ aluminium conductor service cable.

When the means of earthing is the metallic sheath of the supply undertaking's service cable it can be assumed, in most cases, that $Z_E$ will not exceed 0·8 Ω.

Where it is impracticable to comply with the bonding requirements of the PME Approval a PME earthing terminal will not be offered by the supply undertaking and the installation has to be provided with an independent earth electrode. Such an electrode has to be provided for in the design of any installation where the supply undertaking does not offer an earthing terminal.

The resistance of the installation earth electrode is taken as being part of the contribution of the installation impedance to the earth fault loop impedance $Z_s$ and not part of $Z_E$. The latter impedance can be assumed to be a maximum of 21 Ω, of which 20 Ω is the maximum permitted by the PME Approval for the supply undertaking's earth electrode at their distribution transformer, the remaining 1 Ω being a nominal allowance for the internal impedance of the mains and service cables.

The pertinent regulations concerning the installation earth electrode itself are *Regulations 542-10* to *542-15* inclusive, and for information on the design of earth electrodes reference should be made to the Code of Practice CP 1013 'Earthing'. The earthing conductor has to comply with *Regulation 543-1* and, if buried in the soil, with *Regulation 542-16*.

When the means of earthing is an earth electrode at the installation, i.e. when the installation is part of a TT system, in many cases the earth fault loop impedance will be too high for overcurrent protective devices to be used to give protection against indirect contact. When this is the case, *Regulation 471-15* requires the use of either residual current devices or fault-voltage operated devices (i.e. voltage-operated earth leakage circuit breakers) the former being preferred, as stated in *Regulation 471-15* itself and in *Regulation 413-12*. In any event, for socket outlet circuits in such installations a residual current device having a rated residual operating current not exceeding 30 mA must be used, as demanded by *Regulation 471-13*.

The earth fault current will be comparatively low but, as shown in Chapter 4 of this Commentary, such a residual current device will operate within 40 ms with an earth fault of only 250 mA. From consideration of the thermal requirements of *Regulation 543-2*, the minimum cross-sectional area of 2·5 mm$^2$ permitted by *Regulation 543-1* (if mechanical protection is provided) is capable of withstanding 1800A for 40 ms. Thus, in an installation which is part of a TT system, the size of earthing conductor is unlikely to be determined from thermal considerations but from mechanical and corrosion considerations.

## 12.6 Main equipotential bonding conductors

The cross-sectional area of the main equipotential bonding conductor must generally comply with *Regulation 547-2* which requires a minimum of 6mm$^2$ and a maximum not normally exceeding 25mm$^2$.

Where the supply is PME, the requirements of the supply undertaking concerned as to the size of these bonding conductors have to be met.

Table 18 shows the sizes of the earthing conductor and main equipotential bonding conductor required for compliance with the relevant regulations, i.e. *Regulations 542-16, 543-1, 543-3* and *547-2*, where these are copper conductors.

This Table is, in fact, based on the adoption of Table 54F, which avoids the need to apply the equation given in *Regulation 543-2* and in this context it is pointed out that if the designer opts to use that equation, some reduction in the sizes given in Table 18 may be achieved, except where minimum sizes are specifically identified in the regulations mentioned above.

The inclusion of data where the phase conductor is less than 16mm$^2$ is to cater for connections of fire alarm installations, intruder alarm installations and the like although supply undertakings will not provide connections for 'tails' of less than 6mm$^2$ owing to the inadequate short circuit protection afforded by their cut-out fuse, notwithstanding the requirements of *Regulation 473-6*.

Furthermore, where cross-sectional areas are reduced by the application of the equation in *Regulation 543-2*, in no circumstances should they be less than the minimum values indicated in the final column of Table 18 if it is intended to accept the offer of a PME earthing terminal. These minimum values should be confirmed with the supply undertaking concerned as some local circumstances may require larger sizes.

The other protective conductors encountered in the type of installation considered here are circuit conductors and supplementary equipotential bonding conductors but these will be dealt with later.

## 12.7 Prospective short circuit current at the origin

Consideration should now be given to the other characteristic of the supply which is required to be known before one can proceed with the design of the installation, namely the prospective short circuit current at the origin of the installation.

A typical maximum value for this prospective short circuit current is 10 kA based on the supply undertaking's transformer having a rating of 1000kVA, a 25 m length of 300mm² mains cable with aluminium conductors and a 6 m length of 25mm² service cable, also with aluminium conductors.

Here again, the supply undertaking is not under a statutory obligation to guarantee a particular value and will not be able to do so, and when estimating the prospective fault current it must be remembered that the supply undertaking may want to reinforce their network in order to meet the load demands of their consumers. Consequently, whatever the initial prospective short circuit current, it is always possible that the supply undertaking could instal a 1000 kVA transformer adjacent to the premises concerned. The length, and hence the impedance, of the service cable has a marked effect on the prospective short circuit current and it is possible to estimate what the current will be for different lengths of that cable.

Assuming the same transformer, length and size of mains cable as before, and ignoring the reactance of the service cable the following estimated prospective short circuit currents are obtained:

| Length of service cable | Prospective short circuit current |
|---|---|
| m | kA |
| 10 | 7·3 |
| 15 | 5·4 |
| 20 | 4·3 |
| 25 | 3·5 |

If a crossroad service is involved, the service length should be taken to the footpath on the same side of the road as the premises concerned because, when the supply to the area is reinforced, extra mains may be laid and services transferred to the new mains. The supply undertaking will not be able to indicate how likely this might be as they would need to predict the requirements of their customers over a considerable number of years.

**Table 18: Minimum sizes of earthing conductors and main equipotential bonding conductors**
(*Earthing conductor selected in accordance with Table 54F of Chapter 54 and main equipotential bonding selected in accordance with Regulation 547-2*)

| Phase conductor | Nonseparate earthing conductor | Related main equipotential bonding conductor | Separate earthing conductor not buried or buried with corrosion and mechanical protection | Related main equipotential bonding conductor | Separate earthing conductor buried with protection against corrosion but not mechanically protected | Related main equipotential bonding conductor | Separate earthing conductor buried but not corrosion protected | Related main equipotential bonding conductor | Minimum size of bonding conductor to meet PME approval |
|---|---|---|---|---|---|---|---|---|---|
| $mm^2$ | $mm^2$ | $mm^2$ | $mm^2$ | $mm^2$ | $mm^2$ | $mm^2$ | $mm^2$ | $mm^2$ | $mm^2$ |
| 4 | 4 | 6* | 4 | 6* | 16* | 10* | 25* | 16* | 6* |
| 6 | 6 | 6* | 6 | 6* | 16* | 10* | 25* | 16* | 6* |
| 10 | 10 | 6* | 10 | 6* | 16* | 10* | 25* | 16* | 6* |
| 16 | 16 | 10 | 16 | 10 | 16* | 10* | 25* | 16* | 6* |
| 25 | 16 | 10 | 16 | 10 | 16* | 10* | 25* | 16* | 16* |
| 35 | 16 | 10 | 16 | 10 | 16* | 10* | 25* | 16* | 16* |
| 50 | 25 | 16 | 25 | 16 | 25 | 16 | 25* | 16* | 16* |
| 70 | 35 | 25 | 35 | 25 | 35 | 25 | 35 | 25 | 50* |

Note: When the cross-sectional areas of earthing conductor are determined by means of the equation in *Regulation 543-2*, the tabulated values *not* marked with an asterisk may be reduced. The tabulated values which are marked with an asterisk must not be reduced in any circumstance.

In order to comply with the requirements of Chapter 43, Section 473 and Section 533 as regards the provision of overload and short circuit protection in the installations being considered here, it will be usual to instal a consumer's distribution unit which will incorporate the main switch or circuit breaker (double-pole for single-phase) demanded by *Regulation 476-15* and the fuses or miniature circuit breakers controlling the final circuits of the installation as demanded by *Regulation 476-16*. Another pertinent regulation is *Regulation 314-4* which requires that each final circuit shall be connected to a separate way in the consumer unit and that the wiring of each final circuit shall be electrically separate from that of any other final circuit. Modern versions of the consumer unit may also include a residual current device.

The installation designer having ascertained the prospective short circuit current at the origin of the installation has then to verify that the consumer unit he intends to use and all its component devices including the main switch is capable of withstanding that value of prospective short circuit current. He may well have to seek the advice of the manufacturer of the consumer unit on this aspect but it is expected that with the advent of the Fifteenth Edition manufacturers may well indicate in their specifications for such units their suitability or otherwise for use in series with the 60A and 100A BS 1361 fuses at various levels of short circuit current.

Provided the protective devices in the consumer unit have an adequate breaking capacity, the designer can then invoke *Regulation 434-5* which allows him to assume, in general terms, that there is no need for him to carry out the determination of short circuit currents throughout the installation as required by *Regulation 434-2*.

In choosing the consumer unit to use the designer has to take account of Section 314 of which one Regulation, namely *Regulation 314-4*, has already been mentioned, particular attention being taken of the requirement prescribed in item (i) of *Regulation 314-1*. Lighting and power should be handled by separate circuits and it is considered that not all the lighting should be on only one circuit.

## 12.8 Use of residual current devices

At this point a major difference between the Fifteenth Edition and preceding editions must be mentioned and that is the requirement given in *Regulation 471-12*. This regulation requires that if a socket outlet circuit is specifically intended to supply equipment to be used outdoors that circuit has to be protected by means of a residual current device having a rated residual operating current not exceeding 30 mA. For installations where there is no metallic path provided between the earthed neutral point of the source of energy and the installation, i.e. where the installation is part of a TT system, *all* socket outlet circuits (in a household or similar installation) have to be protected by means of just such a device (*Regulation 471-13*.

As stated earlier, some modern consumer units incorporate a residual current device which gives protection to all the final circuits. This arrangement has the disadvantage that an earth fault occurring anywhere in the installation will cause disconnection of all circuits and there is the further disadvantage that nuisance tripping could occur because the leakage currents from healthy equipment could add up to a value approaching the residual operating current of the device, this aspect being dealt with by *Regulation 531-5*.

## 12.9 Final circuits

For household installations, it is to be expected that the designer will choose one or more of the standard circuit arrangements given in Table 5A of Appendix 5 of the Wiring Regulations for the power circuits using BS 1363 accessories and the cooker circuit detailed in Item D of the same Appendix.

In terms of the size of conductors used and the current ratings of the over-current protective devices, the circuits are generally the same as those given in the Fourteenth Edition. The number of socket outlets in the radial circuits detailed in Table 5A of Appendix 5 is not limited but as for all these standard circuits the area served by the circuits is now based on the known or estimated load, subject to a maximum area given in that table.

Appendix 5 also makes it clear that in a correctly designed installation:

(*a*) immersion heaters or permanently connected heating appliances forming part of a comprehensive space heating installation should be supplied by their own separate circuits

(*b*) the loading in kitchens (or laundry rooms) may be such as to require a separate circuit, either ring or radial

(*c*) the conductor sizes given in Table 5A assume an ambient temperature of 30°C and that cables of more than two circuits are not bunched together. If either of these parameters are changed the size of conductor has to be increased and it may not be possible to accommodate these sizes in the terminals of the BS 1363 accessories. This would necessitate a re-examination of the design.

Here it must be emphasised that even when the designer opts to use a standard circuit arrangement from Table 5A he must assure himself that it is adequate for the expected load current it will have to carry, i.e. the designer has to satisfy himself that the diversity used in developing the standard circuit arrangements is equally applicable to the installation he is designing.

As regards item (*c*), particular attention must be drawn here to *Regulation 522-6* which concerns cables run for a significant length in a space to which thermal insulation is to be applied. This is a new regulation and has the effect that if the cable is to be partly surrounded by thermal insulation the size of cable given in Table 5A has to be increased to the next size up, or if a cable is to be wholly surrounded by such insulation the size given in Table 5A has to be increased by two sizes. Here again there may be problems in relation to accommodation of the cables in the terminals of accessories and as a general rule it is advisable to avoid cable runs in thermal insulation or in locations where it is likely that such insulation may be added at a later date.

Whether or not the designer decides to use the standard circuit arrangements in Table 5A each final circuit has to be such that:

(i) the earth fault loop impedance does not exceed the value tabulated in Tables 41A1 or Table 41A2 appropriate to the type and current rating of the overcurrent protective device associated with the circuit *but only if it is intended that the device is also to give protection against indirect contact.* If that protection is to be given by a residual current device the earth fault loop impedance (in ohms) multiplied by the rated residual operating current (in amperes) must not exceed 50.

(ii) the earth fault loop impedance must be such that the circuit protective conductor complies with the thermal requirements of *Regulations 543-2.*

Tables 8A to 8C of Appendix 8 give values of earth fault loop impedance applicable equally to ring and radial circuits which, if not exceeded, give compliance with both items (i) and (ii) above.

There is a third requirement all final circuits have to meet, namely,

(iii) the resistances of the live conductors (i.e. the phase and neutral conductors) must be such that the voltage drop under normal load conditions does not exceed 2·5% of the nominal supply voltage, as demanded by *Regulation 522-8.*

It will be found that in many cases it is this third item which limits the length of the circuit, not the requirements concerning the earth fault loop impedance or the thermal requirement associated with short circuit currents. Resistance has been used in item (iii) as it is assumed that the cables used will have cross-sectional areas less than 35mm$^2$ so that their inductances can be ignored.

Every circuit, other than a standard circuit arrangement given in Table 5A of Appendix 5 and with certain exceptions given in the Wiring Regulations, has to be designed so as to comply with *Regulation 433-1* as regards overload protection and with *Regulation 434-1* as regards short circuit protection. For the latter, as already indicated, provided that the overcurrent protective devices in the consumer unit have a breaking capacity not less than the prospective short circuit current, the designer does not need to check compliance with *Regulation 434-6*.

Thus, in designing a final circuit the following sequence of steps suggests itself.

The first stage is to determine its design current or maximum demand ($I_B$ A) and to do this it is necessary to determine the total current demand ($I_T$ A) of the points of utilisation and current-using equipment the circuit is intended to serve.

The current demands to be assumed for the individual points and equipment are given in Table 4A of Appendix 4 of the Wiring Regulations and the designer is then allowed by *Regulation 311-2* to apply an allowance for diversity, denoted here by $\alpha$, to $I_T$ in order to obtain the design current of the circuit, i.e.

$$I_B = \alpha I_T \text{ A}$$

Table 4B of Appendix 4 gives values of $\alpha$ which may be used for the types of installation considered here. If for a particular installation there is any doubt as to the value of $\alpha$ which is applicable, it becomes necessary to use a value sufficiently high to give a margin of safety. If too optimistic a value is used, this could lead to a prolonged or even continuous overload to occur of insufficient magnitude to cause operation of the overcurrent protective device of the circuit but sufficient to cause some degree of overheating of the circuit conductors which, at the least, could lead to accelerated deterioration of the cable insulation.

This first stage will also indicate to the designer the best subdivision of the installation into the necessary number of final circuits, not only from the aspect of division of the load as such but also to meet the requirements of *Regulation 314-1*.

Assuming that the type of cable and the method of installation of that cable have already been chosen, account having been taken of the environmental conditions likely to be encountered so that the relevant requirements of Section 523 of the Wiring Regulations are met, the next stage in the circuit design is to determine the minimum cross-sectional area of the cable conductors which can be used.

In order to do this the designer however must first determine the nominal current rating $I_n$ of the overcurrent protective device it is intended to use. In doing this the designer must bear in mind the requirement prescribed in item (i) of *Regulation 433-2* that this nominal current must not be less than the design current $I_B$ of the circuit. Having chosen $I_n$ the designer then proceeds as follows.

Where the overcurrent protective device is other than a semi-enclosed fuse to BS 3036 the cable has to be such that its tabulated current-carrying capacity (appropriate to the method of installation adopted) has to be not less than

$$I_n \times \frac{1}{C_1} \times \frac{1}{C_2} \times \frac{1}{C_3} \text{ A}$$

where

$C_1$ = correction factor for grouping
$C_2$ = correction factor for ambient temperature
$C_3$ = correction factor if cable is installed in contact with, or surrounded by, thermal insulation.

Where the overcurrent protective device is a semi-enclosed fuse to BS 3036 a further correction factor has to be introduced so that in this case the tabulated current-carrying capacity of the cable it is hoped to use has to be not less than

$$I_n \times \frac{1}{C_1} \times \frac{1}{C_2} \times \frac{1}{C_3} \times \frac{1}{0{\cdot}725}\ \text{A}$$

The factor 1/0·725 is *not* used if the cables are mineral-insulated cables.

There may well be other factors influencing the cross-sectional area of circuit conductor which can be used; for example, if the circuit is supplying a motor load subject to frequent stopping and starting this may demand a larger size of cable.

At this stage the design steps that have been taken are such that the circuit complies with the requirements concerning protection against overload current, as prescribed in *Regulation 433-2.*

The next aspect to be considered is the requirement which is applicable to all circuits, namely the limitation of voltage drop under normal load conditions, as prescribed by *Regulation 522-8,* this voltage drop being required not to exceed 2½% of the nominal voltage of the supply.

Whether the circuit conductors determined, as already explained, from consideration of the requirements concerning overload protection also comply with the 2½% limitation in voltage drop is readily checked using the voltage drop column of the relevant table in Appendix 9 of the Wiring Regulations.

However, one of the major problems encountered in respect of the determination of voltage drop in a circuit and checking that it complies with the 2½% limitation concerns multi-socket outlet circuits, including the standard circuit arrangements detailed in Table 5A of Appendix 5 of the Wiring Regulations.

In some industrial and commercial installations where a multi-socket outlet circuit feeds a number of items of current-using equipment, for instance, a bank of sewing machines or, in the latter, a number of typewriters, the assumption can be made that all the items may be in operation at the same time. Estimating the distances, and hence the resistance of the conductors between the socket outlets the designer (provided he has knowledge of the type and current ratings of the current-using equipment) is then able to calculate the maximum voltage drop.

In a household installation, such an approach is obviously not possible, for the simple reason that there is no way of predetermining the appliances which will be fed from the socket outlets concerned. In these circumstances one is forced to adopt a hypothetical basis for design and a minimum estimate of the loading would be that of a 3kW appliance fed from the most distant socket outlet of a multi-socket outlet circuit.

A possible approach would be to consider the probability of two 3kW appliances being fed from the most distant socket outlets of the circuit concerned as it is now common practice to fit twin-socket outlets rather than single outlets. A statement in Appendix 5 of the Wiring Regulations has a bearing on the probable loading, indicating as it does that consideration should be given to the loading in kitchens.

It must be emphasised that the following analysis based on the above assumption does not mean that in practice one can ignore the possibility of a circuit being loaded to the current rating of the overcurrent protective device. The main aim of this particular analysis is to compare the possible maximum lengths of the standard circuits based on voltage drop as compared with the limitations imposed by other requirements.

The restriction imposed on the length of circuit necessary to meet the 2½% limitation in voltage drop can then be compared with the restriction imposed by the need to meet the maximum values of earth fault loop impedance in order to comply with the maximum disconnection time of 0·4 s specified in *Regulation 413-4*. The results of this comparison will indicate if there is any possibility of relaxing the assumed loading conditions, in which case a more careful assessment would be needed of the probable maximum loading. Adopting the more prudent approach, the maximum resistance of the phase conductor of a single phase radial circuit having a nominal voltage of 240V is then 3/26 Ω, i.e. 0·115 Ω.

Similarly, the effective resistance of the phase conductor of a 240 V single-phase ring circuit must be 0·115 Ω. This is made up of two conductors in parallel each having a resistance of 0·23 Ω or, in other words, a maximum total resistance of 0·46 Ω when measured before the ring is completed.
Now consider the following circuits from Table 5A of Appendix 5 of the Wiring Regulations.

*Circuit A1:* This is a ring circuit and assume that the cable used is the 2·5mm² 'twin-and-earth' p.v.c.-insulated cable to BS 6004 having a 1·5mm² protective conductor. The maximum length of the circuit for compliance with the 2½% voltage drop limitation is 0·46/0·0085 m, i.e. 54 m, 0·0085 Ω being the resistance of a 2·5mm² conductor at 70°C, the maximum permissible temperature for p.v.c.
*Circuit A2:* This is a radial circuit and assume that the cable used is the 4mm² 'twin-and-earth' p.v.c.-insulated cable to BS 6004 having a 1·5mm² protective conductor. The maximum length of this circuit for compliance with the 2½% voltage drop limitation is 0·115/0·005 m, i.e. 21 m.

The above lengths of circuits have now to be compared with the maximum lengths imposed by the requirements concerning, for instance, protection against indirect contact.

Table 8D of Appendix 8 of the Wiring Regulations gives the combined resistances of the phase and protective conductors for p.v.c.-insulated cables to BS 6004 which assume that under earth fault conditions the conductors reach the maximum permissible temperature given in Chapter 54, namely 160°C.

Thus the circuits just described have a combined resistance of phase and protective conductor (i.e. $R_1 + R_2$) of

| | |
|---|---|
| for circuit A1 | 54 x 0·03 Ω = 1·62 Ω |
| for circuit A2 | 21 x 0·026 Ω = 0·55 Ω |

As circuit A1 is a ring circuit, it is more convenient to use its effective resistance (see Appendix D of this Commentary) and this is 1·62/4 Ω, i.e. 0·41 Ω.

Comparison between these values and those given in the appropriate tables in Appendix 8 of the Wiring Regulations for the maximum values of earth fault loop impedance $Z_s$ for compliance with both the 0·4 s disconnection time demanded by *Regulation 413-4* and the thermal requirement of *Regulation 543-2* gives the following results:

*Circuit A1:* Table 5A of Appendix 5 has no restriction as to the type of overcurrent protective device used but the current rating has to be 30A or 32A. From Table 8A and Table 8B of Appendix 8 and remembering that the protective conductor is 1·5mm$^2$ in cross-sectional area, the earth fault loop impedance must not exceed 1·1 Ω for a BS 3036 semi-enclosed fuse (30A) and a BS 88 Part 2 fuse (32A), respectively.

Deducting 0·41 Ω, 0·69 Ω is left for the earth loop impedance external to the installation $Z_E$.

If a 30A fuse to BS 1361 is used, and this is the most common HBC fuse in domestic premises, then (1·2 − 0·41) Ω, i.e. 0·79 Ω, is left for the value of $Z_E$.

If a 30A Type 1 miniature circuit breaker is used, the maximum earth fault loop impedance is obtained from Table 41A1 of Chapter 41 of the Wiring Regulations and is found to be 2 Ω, and therefore in this case $Z_E$ could be 1·69 Ω. If a Type 2 miniature circuit breaker is used, $Z_E$ must not exceed 0·74 Ω.

The implications are therefore that such a circuit using any of these overcurrent protective devices can be used in an installation fitted with a PME earthing terminal where it has been confirmed that $Z_E$ is unlikely to exceed 0·35 Ω. If the means of earthing is the lead sheath of the supply undertaking's service cable, where the maximum expected $Z_E$ is 0·8 Ω, then a Type 1 miniature circuit breaker, or a Type 2 miniature circuit breaker, or a BS 1361 30A fuse with a minor reduction in the length of the circuit, would be appropriate.

It can be seen from the above analysis that the major restriction as regards the permissible length of the A1 circuit, particularly when a PME earthing terminal is available, is the volt drop requirement. Consequently, should the designer of the installation find that the permitted length of circuit too short, he has the option of making a more careful estimate of the likely use of appliances on the ring circuit and, if he is then able to show that the assumption of two 3kW appliances at the midpoint of the ring circuit can be relaxed, the length of the circuit can be extended up to the lengths implied by the need to meet the limitation of earth fault loop impedance.

If the designer cannot relax the assumption made on loading but still wishes to extend the circuit, then he must either use a cable having a greater cross-sectional area or specify a residual current device; if the second course is chosen, overcurrent protection must still be provided for the circuit and it must be confirmed that the increase in circuit length does not increase the phase-neutral impedance loop sufficiently to nullify that protection.

*Circuit A2:* For this radial circuit, Table 5A in Appendix 5 does not allow a semi-enclosed fuse to BS 3036 to be used, so it is only necessary to consider HBC fuses and miniature circuit breakers. As before, by using the limiting values of earth fault loop impedance $Z_s$ from the appropriate table in Appendix 8 and, for the miniature circuit breaker, from Table 41A1 of Chapter 41, it is readily established, (remembering that the maximum combined resistance of the circuit phase and protective conductors when the circuit complies with the 2½% drop limitation is 0·55 Ω) that the maximum values of $Z_E$ are:

When a BS 88 Part 2 fuse is used

$$Z_E = (1{\cdot}1 - 0{\cdot}55)\ \Omega = 0{\cdot}55\ \Omega$$

When a BS 1361 fuse is used

$$Z_E = (1{\cdot}2 - 0{\cdot}55)\ \Omega = 0{\cdot}65\ \Omega$$

When a Type 1 miniature circuit breaker is used

$$Z_E = (2 - 0{\cdot}55)\ \Omega = 1{\cdot}45\ \Omega$$

When a Type 2 miniature circuit breaker is used

$$Z_E = (1{\cdot}15 - 0{\cdot}55)\ \Omega = 0{\cdot}6\ \Omega$$

When a Type 3 miniature circuit breaker is used

$$Z_E = (0{\cdot}8 - 0{\cdot}55)\ \Omega = 0{\cdot}25\ \Omega.$$

Clearly all the overcurrent protective devices, other than the Type 3 miniature circuit breaker, will be satisfactory when used in conjunction with a PME earthing terminal where, as before, it has been confirmed that $Z_E$ is unlikely to exceed 0·35 Ω. If the means of earthing is the lead sheath of the supply undertaking's service cable then, if it is desired to use HBC fuses it may be worth investigation, in a particular area, of the likelihood of the earth fault loop impedance external to the installation ($Z_E$) not exceeding 0·55 Ω or 0·65 Ω, as appropriate, or, if one wishes to use Type 2 miniature circuit breakers, that $Z_E$ would not exceed 0·6 Ω. If it is not possible to obtain assurances that these values of $Z_E$ would not be exceeded, it becomes necessary, if one wishes to retain the length of circuit of 21 m, either to increase the conductor cross-sectional area or use a Type 1 miniature circuit breaker.

*Circuit A3.* This is the 20A radial circuit detailed in Table 5A of Appendix 5 of the Wiring Regulations. If a similar analysis is made for this circuit, then by assuming the worst possible loading condition of 20A taken from a twin-socket outlet at the remote end of the circuit, the resistance of the phase conductor, from voltage drop considerations must not exceed 3/20 Ω, i.e. 0·15 Ω.

The maximum length of circuit is therefore 0·15/0·0085 m, i.e. 17·6 m, and this in turn leads to a combined phase and protective conductor resistance (at 160°C) of 17·6 x 0·03 Ω, i.e. 0·53 Ω, assuming of course that the cable used in 2·5mm² 'twin-and-earth' p.v.c.-insulated cable to BS 6004 with a 1·5mm² protective conductor.

From the appropriate tables in Appendix 8, it will be seen that irrespective of the type of fuse the maximum earth fault loop impedance $Z_s$ is 1·8 Ω and from Table 41A1 this impedance is, for Type 1 m.c.b.s, 3 Ω; for Type 2 m.c.b.s, 1·7 Ω;

and for Type 3 m.c.b.s, 1·2 Ω.

Thus, for all protective devices there is no difficulty in meeting these limiting values of $Z_s$ where a PME earth terminal is offered and, with the exception of the Type 3 m.c.b, the same is true where the means of earthing is the lead sheath of the supply undertaking's service cable.

A detailed examination of the three standard circuit arrangements would show that in the great majority of cases it is the normal voltage drop limitation which determines the maximum length of circuit.

Table 19 therefore assumes the designer wishes to run the circuit concerned for the maximum length as determined earlier and gives the maximum earth fault loop impedance $Z_s$ for compliance with *Regulations 413-4* and *543-2*. From these two parameters the maximum value of $Z_E$, that part of the earth loop impedance external to the installation, is readily determined, the values being as given in the table.

Thus, from Table 19 it will be seen immediately that if a PME terminal is offered and can be used, as the maximum value of $Z_E$ can be taken to be 0·35 Ω, all the circuits with two exceptions related to the Type 3 miniature circuit breaker could be employed.

When the earthing arrangement is a terminal connected to the lead sheath of the supply undertaking's service cables $Z_E$, as already stated, may be 0·8 Ω, and in this case Table 19 indicates that a number of the circuits would not be suitable. Unless it could be shown, for a particular installation, that $Z_E$ will not exceed the tabulated value, it becomes necessary to either reduce the circuit length or increase the conductor size.

An alternative approach to the determination of the maximum length of a ring circuit for compliance with the voltage drop limitation would be to assume a number of socket outlets are equally loaded and equally spaced around that circuit.

For instance, if it is assumed that the standard ring circuit A1 has fifteen socket outlet positions equally spaced around it and each socket outlet is supplying a load of 2A, then it can be readily shown that the maximum length of the ring circuit for compliance with the voltage drop requirement is 88 m.

The number of possible socket outlet arrangements and loadings is, of course, infinite, and includes the combination, for instance, of lightly loaded socket outlets with one or two socket outlets loaded to the maximum permitted 13A. An analysis of some of the possible arrangements suggests that the maximum length of the standard ring circuit can be between 45 and 90 m.

Tables similar to that shown for the standard circuit arrangements are readily prepared for other circuits, the following being typical examples. Tables 20 and 21 deal with BS 88 fuses and Tables 22 and 23 deal with BS 3036 fuses. For each type of fuse the tables cover socket outlet circuits and circuits supplying fixed equipment.

Finally, in this section dealing with final circuits it is necessary to consider their compliance with the thermal requirements of *Regulation 434-6* when subjected to short circuit currents.

There are two aspects that must be considered. First, the overcurrent device protecting the final circuit has to be capable of withstanding the prospective short circuit current at the point where that device is installed, taking account of the characteristic of the overcurrent protective device on the supply side of that point.

Provided that the protective devices in the consumer unit have an adequate breaking capacity, the designer is then able to assume, as permitted by *Regulation 434-5*, that there is no need to carry out the determination of short circuit currents throughout the installation as required by *Regulation 434-2* or to check the compliance of the final circuits with the thermal requirement prescribed in *Regulation 434-6*.

When this assumption cannot be made, the overcurrent protective devices in the consumer unit must have a sufficiently rapid disconnection time to protect the conductors of the final circuits against an excessive temperature rise when the short circuit fault occurs at the remote end of the circuits. This is the second aspect referred to earlier.

Thus, when it is necessary to check compliance with *Regulation 434-6*, this *minimum* short circuit current has to be determined because this is the most onerous condition so far as meeting the requirements of this regulation. It is understood that the minimum prospective short circuit current at the origin of an installation fed from the public supply network is unlikely to be less than 600A and it can be shown that even at this level of short circuit current the circuits detailed in the previous table will also comply with *Regulation 434-6*.

With a minimum short circuit current at the origin of the installation of 600A, then for a 240V single-phase circuit this gives a phase-neutral impedance $Z_C$ external to the installation of 240/600 Ω, i.e. 0·4 Ω. As already stated, for the standard circuit arrangements A1, A2 and A3, the limiting factor is generally the 2·5% voltage drop limitation so that for these the maximum circuit lengths are 54 m. 21 m and 17·6 m respectively.

The combined phase plus neutral conductor resistances for these maximum lengths being:

for circuit A1 : 0·23 Ω
for circuit A2 : 0·23 Ω
for circuit A3 : 0·3 Ω

The maximum short-circuit current for each circuit is therefore

$$\text{for circuit A1} = \frac{240}{0{\cdot}4 + 0{\cdot}23}\ \text{A} = 380\text{A}$$

$$\text{for circuit A2} = \frac{240}{0{\cdot}4 + 0{\cdot}23}\ \text{A} = 380\text{A}$$

$$\text{for circuit A3} = \frac{240}{0{\cdot}4 + 0{\cdot}3}\ \text{A} = 343\text{A}$$

Examination of the graphs given in Chapter 6 of this Commentary shows that in all cases, irrespective of the type of overcurrent protective device used, the circuits also

**Table 19: Maximum values of $Z_S$ and $Z_E$ for the standard circuit arrangements of Table 5A of Appendix 5 of the Wiring Regulations**

| Type of circuit to Table 5A | Cross-sectional area of phase conductor | Max. length | Current rating and type of overcurrent protective device | When using 'twin-and-earth' cables to BS 6004 (i.e. with 1.5mm² protective conductor) | | When using single-core cables to BS. 6004 (i.e. protective conductor same size as phase conductor) | |
|---|---|---|---|---|---|---|---|
| | mm² | m | | max $Z_s$ Ω | max $Z_E$ Ω | max $Z_s$ Ω | max $Z_E$ Ω |
| | | | 30A. BS 3036 Fuse | 1.1 | 0.69 | 1.1 | 0.8 |
| | | | 32A. BS 88 Pt 2 Fuse | 1.1 | 0.69 | 1.1 | 0.8 |
| $A_1$ | 2.5 | 54 | 30A. BS 1361 Fuse | 1.2 | 0.79 | 1.2 | 0.9 |
| (ring) | | | 30A Type 1 m.c.b | 2.0 | 1.59 | 2.0 | 1.7 |
| | | | 30A Type 2 m.c.b | 1.14 | 0.73 | 1.14 | 0.84 |
| | | | 30A Type 3 m.c.b | 0.8 | 0.39 | 0.8 | 0.5 |
| | | | 32A BS 3036 Fuse | not permitted | | | |
| | | | 32A BS 88 Pt 2 Fuse | 1.1 | 0.55 | 1.1 | 0.64 |
| $A_2$ | 4.0 | 21 | 30A BS 1361 Fuse | 1.2 | 0.65 | 1.2 | 0.74 |
| (radial) | | | 30A Type 1 m.c.b | 2.0 | 1.45 | 2.0 | 1.54 |
| | | | 30A Type 2 m.c.b. | 1.14 | 0.59 | 1.14 | 0.68 |
| | | | 30A Type 3 m.c.b. | 0.8 | 0.25 | 0.8 | 0.34 |
| | | | 20A BS 3036 Fuse | 1.8 | 1.27 | 1.8 | 1.4 |
| | | | 20A BS 88 Pt 2 Fuse | 1.8 | 1.27 | 1.8 | 1.4 |
| $A_3$ | 2.5 | 17.6 | 20A BS 1361 Fuse | 1.8 | 1.27 | 1.8 | 1.4 |
| (radial) | | | 20A Type 1 m.c.b. | 3.0 | 2.47 | 3.0 | 2.6 |
| | | | 20A Type 2 m.c.b. | 1.7 | 1.17 | 1.7 | 1.3 |
| | | | 20A Type 3 m.c.b. | 1.2 | 0.67 | 1.2 | 0.8 |

**Table 20: Maximum length and maximum $Z_E$ for 240V single-phase circuits feeding socket outlets. Circuits protected by BS 88 Pt. 2 fuses, copper conductors.**

| Cross-sectional area mm² | | Current rating of BS 88 PT. 2 fuse | | | | | | | | | | | |
|---|---|---|---|---|---|---|---|---|---|---|---|---|---|
| | | 16A | | 20A | | 25A | | 32A | | 40A | | 50A | |
| Phase conductor | Protective conductor | Max length m | Max $Z_E$ Ω | Max length m | Max $Z_E$ Ω | Max length m | Max $Z_E$ Ω | Max length m | Max $Z_E$ Ω | Max length m | Max $Z_E$ Ω | Max length m | Max $Z_E$ Ω |
| 1.0 | 1.0 | 8.9 | 2.31 | | | | | | | | | | |
| 1.5 | 1.0 | 13.2 | 2.19 | 10.5 | 1.32 | | | | | | | | |
| | 1.5 | 13.2 | 2.31 | 10.5 | 1.41 | | | | | | | | |
| 2.5 | 1.5 | 22.0 | 2.13 | 17.6 | 1.27 | 14.1 | 1.08 | | | | | | |
| | 2.5 | 22.0 | 2.32 | 17.6 | 1.41 | 14.1 | 1.19 | | | | | | |
| 4 | 1.5 | 34.0 | 2.12 | 27.3 | 1.09 | 21.8 | 0.93 | 17.0 | 0.66 | | | | |
| | 4 | 34.0 | 2.32 | 27.3 | 1.42 | 21.8 | 1.19 | 17.0 | 0.86 | | | | |
| 6 | 2.5 | 52.8 | 2.15 | 42.3 | 1.12 | 33.8 | 0.96 | 26.4 | 0.68 | 21.1 | 0.46 | | |
| | 6 | 52.8 | 2.31 | 42.3 | 1.41 | 33.8 | 1.19 | 26.4 | 0.86 | 21.1 | 0.61 | | |
| 10 | 4 | | | | | 57.1 | 0.94 | 44.6 | 0.66 | 35.7 | 0.45 | 28.6 | 0.32 |
| | 10 | | | | | 57.1 | 1.19 | 44.6 | 0.85 | 35.7 | 0.60 | 28.6 | 0.44 |
| 16 | 6 | | | | | | | 69.4 | 0.66 | 55.6 | 0.52 | 44.4 | 0.32 |
| | 16 | | | | | | | 69.4 | 0.86 | 55.6 | 0.64 | 44.4 | 0.44 |

Note: The assumption has been made that the design current of the circuit equals the rated current of the protective device.

**Table 21: Maximum length and maximum $Z_E$ for 240V single-phase circuits feeding fixed equipment. Circuits protected by BS 88 PT.2 fuses, copper conductors**

| Cross-sectional area $mm^2$ | | Current rating of BS 88 PT 2 fuse | | | | | | | | | | | | | | | |
|---|---|---|---|---|---|---|---|---|---|---|---|---|---|---|---|---|---|
| | | 6A | | 10A | | 16A | | 20A | | 25A | | 32A | | 40A | | 50A | |
| Phase conductor | Protective conductor | Max length m | Max $Z_E$ Ω | Max length m | Max $Z_E$ Ω | Max length m | Max $Z_E$ Ω | Max length m | Max $Z_E$ Ω | Max length m | Max $Z_E$ Ω | Max length m | Max $Z_E$ Ω | Max length m | Max $Z_E$ Ω | Max length m | Max $Z_E$ Ω |
| 1.0 | 1.0 | 23.8 | 10 | 14.3 | 6.9 | 8.9 | 3.91 | | | | | | | | | | |
| 1.5 | 1.0 | 35.2 | 11.7 | 21.1 | 6.7 | 13.2 | 3.79 | 10.5 | 1.72 | | | | | | | | |
| | 1.5 | 35.2 | 11.9 | 21.1 | 6.9 | 13.2 | 3.91 | 10.5 | 2.51 | | | | | | | | |
| 2.5 | 1.5 | 58.8 | 11.4 | 35.3 | 6.6 | 22.0 | 3.73 | 17.6 | 2.37 | 14.1 | 1.58 | | | | | | |
| | 2.5 | 58.8 | 11.9 | 35.3 | 6.9 | 22.0 | 3.92 | 17.6 | 2.61 | 14.1 | 2.09 | | | | | | |
| 4 | 1.5 | | | | | 34.0 | 3.52 | 27.3 | 2.18 | 21.8 | 1.43 | 17.0 | 0.96 | | | | |
| | 4 | | | | | 34.0 | 3.92 | 27.3 | 2.62 | 21.8 | 2.09 | 17.0 | 1.76 | | | | |
| 6 | 2.5 | | | | | 52.8 | 3.55 | 42.3 | 2.32 | 33.8 | 1.86 | 26.4 | 1.58 | 21.1 | 0.76 | | |
| | 6 | | | | | 52.8 | 3.91 | 42.3 | 2.61 | 33.8 | 2.09 | 26.4 | 1.76 | 21.1 | 1.21 | | |
| 10 | 4 | | | | | | | | | 57.1 | 1.84 | 44.6 | 1.56 | 35.7 | 1.05 | 28.6 | 0.82 |
| | 10 | | | | | | | | | 57.1 | 2.09 | 44.6 | 1.75 | 35.7 | 1.20 | 28.6 | 0.94 |
| 16 | 6 | | | | | | | | | | | 69.4 | 1.56 | 55.6 | 1.12 | 44.4 | 0.82 |
| | 16 | | | | | | | | | | | 69.4 | 1.76 | 55.6 | 1.34 | 4.4 | 0.94 |

Note: The assumption has been made that the design current of the circuit equals the rated current of the protective device.

**Table 22: Maximum length and maximum $Z_E$ for 240V single-phase circuits feeding socket outlets. Circuits protected by BS 3036 fuses, copper conductors**

| Cross-sectional area mm² | | Current rating of BS 3036 fuse | | | | | | | | | |
|---|---|---|---|---|---|---|---|---|---|---|---|
| | | 5A | | 15A | | 20A | | 30A | | 45A | |
| Phase conductor | Protective conductor | Max length m | Max $Z_E$ Ω | Max length m | Max $Z_E$ Ω | Max length m | Max $Z_E$ Ω | Max length m | Max $Z_E$ Ω | Max length m | Max $Z_E$ Ω |
| 1.0 | 1.0 | 28.5 | 8.04 | | | | | | | | |
| 1.5 | 1.0 | 42.2 | 7.66 | | | | | | | | |
| | 1.5 | 42.2 | 8.04 | | | | | | | | |
| 2.5 | 1.5 | 70.5 | 7.48 | 23.5 | 2.0 | 17.6 | 1.27 | | | | |
| | 2.5 | 70.5 | 8.05 | 23.5 | 2.18 | 17.6 | 1.41 | | | | |
| 4 | 1.5 | | | 36.3 | 1.76 | 27.2 | 1.09 | | | | |
| | 4 | | | 36.3 | 2.19 | 27.2 | 1.42 | | | | |
| 6 | 2.5 | | | | | 42.2 | 1.12 | 28.1 | 0.65 | | |
| | 6 | | | | | 42.2 | 1.41 | 28.1 | 0.84 | | |
| 10 | 4 | | | | | | | 47.6 | 0.63 | 31.7 | 0.29 |
| | 10 | | | | | | | 47.6 | 0.84 | 31.7 | 0.43 |
| 16 | 6 | | | | | | | 74.0 | 0.63 | 49.3 | 0.28 |
| | 16 | | | | | | | 74.0 | 0.84 | 49.3 | 0.43 |

Note. The assumption has been made that the design current of the circuit equals the rated current of the protective device.

**Table 23: Maximum length and maximum $Z_E$ for 240V single-phase circuits feeding fixed equipment. Circuits protected by BS 3036 fuses, copper conductors**

| Cross-sectional area mm² | | Current rating of BS 3036 fuse | | | | | | | | | |
|---|---|---|---|---|---|---|---|---|---|---|---|
| | | 5A | | 15A | | 20A | | 30A | | 45A | |
| Phase conductor | Protective conductor | Max length m | Max $Z_E$ Ω | Max length m | Max $Z_E$ Ω | Max length m | Max $Z_E$ Ω | Max length m | Max $Z_E$ Ω | Max length m | Max $Z_E$ Ω |
| 1.0 | 1.0 | 28.5 | 18.4 | | | | | | | | |
| 1.5 | 1.0 | 42.2 | 18.0 | | | | | | | | |
| | 1.5 | 42.2 | 18.4 | | | | | | | | |
| 2.5 | 1.5 | 70.5 | 17.8 | 23.5 | 4.89 | 17.6 | 3.47 | | | | |
| | 2.5 | 70.5 | 18.4 | 23.5 | 5.08 | 17.6 | 3.61 | | | | |
| 4 | 1.5 | | | 36.3 | 4.66 | 27.2 | 3.29 | | | | |
| | 4 | | | 36.3 | 5.09 | 27.2 | 3.62 | | | | |
| 6 | 2.5 | | | | | 42.2 | 3.32 | 28.1 | 2.35 | | |
| | 6 | | | | | 42.2 | 3.61 | 28.1 | 2.54 | | |
| 10 | 4 | | | | | | | 47.6 | 2.33 | 31.7 | 1.29 |
| | 10 | | | | | | | 47.6 | 2.54 | 31.7 | 1.43 |
| 16 | 6 | | | | | | | 74.0 | 2.33 | 49.3 | 1.28 |
| | 16 | | | | | | | 74.0 | 2.54 | 49.3 | 1.43 |

Note. The assumption has been made that the design current of the circuit equals the rated current of the protective device.

comply with *Regulation 434-6*.

The foregoing analysis has been based on the use of overcurrent protective devices to give also protection against indirect contact. When it is necessary to use a residual current device to give that protection, as demanded by *Regulations 471-12* and *471-13*, or to use a fault-voltage operated protective device, the design approach to be adopted depends on whether the installation concerned, or more precisely the circuits in that installation, together with the source of energy, constitute one of the family of TN systems or a TT system.

In either case, the circuits, of course, have to comply with the voltage drop limitation under normal load conditions and they have to be protected against overload and short circuit currents as already explained. As has been pointed out, for the installations considered here and where the circuits are the standard circuit arrangements of Table 5A of Appendix 5 of the Wiring Regulations, in general terms the limitation of voltage drop usually determines the length of circuit so that the maximum lengths suggested by the foregoing Tables still apply.

The cross-sectional area of the circuit protective conductor, when the circuit is part of a TN system, i.e. when that conductor is connected to a PME earthing terminal or to the lead sheath of the supply undertaking's service mains (in either case via the earthing conductor), is determined in exactly the same way as already described when protection against indirect contact is provided by an overcurrent protective device. In other words the designer can opt to use Table 54F or determine the minimum cross-sectional area for compliance with the adiabatic equation given in *Regulation 543-2*, subject to the minimum values prescribed in *Regulation 543-1*.

As already explained in Chapter 4 of this Commentary, the earth fault current, still limiting consideration to a circuit which is part of a TN system, will be many times the rated residual operating current of the device and the operating time of the device is going to be less than 0·04 s; in fact, the disconnection time will approach 0·01 s. In order to check that the circuit complies with the thermal requirements of *Regulation 543-2*, it is now necessary to determine the maximum earth fault current which can flow, corresponding to the earth fault occurring near the origin of the circuit. Thus, when it is intended to use a residual current device in an installation which is part of a TN system it is necessary to have some indication of the minimum earth fault loop impedance external to the installation in addition to its maximum value.

If it is intended to use, for instance, p.v.c.-insulated twin-and-earth cables to BS 6004 in one of the standard circuit arrangements to Table 5A of Appendix 5 the cross-sectional area of the protective conductor is 1·5mm$^2$, and assuming the disconnection time is 0·01 s, the fault current which can be tolerated from the adiabatic equation in *Regulation 543-2* is approximately 1700A. This suggests that the minimum earth fault loop impedance external to the installation should not be less than 0·14 Ω (when $U_0$ = 240 V) but it must be remembered that the overcurrent device protecting the circuit, if for instance a 30A fuse to BS 1361 or a 32A fuse to BS 88 Part 2 will exhibit a cut-off current of the order of 650A. It will be appreciated that the residual current device must be capable of withstanding the level of fault energy associated with the above value of cut-off current, and this also applies under short circuit conditions. It is emphasised that, in theory, it is necessary to calculate the maximum earth fault loop impedance of a circuit, corresponding to the earth fault occurring at the remote end of that circuit, in order to check that the product of this impedance and the rated residual operating current of the device does not exceed 50, as demanded by *Regulation 413-6*. However, when the device has a rated residual operating current of 30mA or less and the installation is part of a TN system this calculation is not normally necessary.

When the circuit is part of a TT system, i.e. its protective conductor is connected via an earthing conductor to an earth electrode at the installation, the earth fault currents will be much lower than for the TN case just described but again will usually be such as to give disconnection in 0·04s or less. When the nominal voltage to Earth is 240V, and an r.c.c.b having a rated residual operating current of 30mA is used in order to attain this disconnection time, the earth fault current is required to be at least 0·25A, in which case the earth fault loop impedance must not exceed 960 Ω. Should the rated residual operating current be higher the maximum permitted value of this impedance is correspondingly reduced but, in practice, the appropriate values should be readily achieved.

It follows that the circuit protective conductors in this case can be considerably less than those in installations which are part of a TN system although on the basis of convenience it can be argued that the use of standard twin-and-earth cables is still justifiable although the cross-sectional area of the circuit protective conductor would then be greater than it need be in order to meet the requirements of *Regulations 543-1* and *543-2*. However, it should also be borne in mind that an installation initially part of a TT system may at some later date be offered a PME earthing terminal and this is another reason why it is advisable to use the size of protective conductor based on the installation being part on TN system in order at the later date to be able to take advantage of that earthing terminal.

## 12.10 Supplementary equipotential bonding conductors

The basic requirement concerning supplementary equipotential bonding is prescribed in *Regulation 413-7*. This has the effect of demanding such bonding to be carried out where extraneous conductive parts in the zone concerned (i.e. that formed by the *main* equipotential bonding) are simultaneously accessible with other conductive parts (extraneous or exposed) and where at the same time there is some doubt as to the permanency and reliability of metal-to-metal joints in the extraneous conductive parts. Here attention is drawn to the definition of 'extraneous conductive part' which makes it clear that the term relates only to parts liable to transmit a potential, including earth potential.

The simplest example that one can use to illustrate this aspect is that of a metal sink unit. If there are permanent and reliable metal-to-metal joints between the taps and sink and between the waste pipe and sink, then provided that there are similar joints throughout the cold and hot water pipe systems there is no need to carry out supplementary equipotential bonding.

It is extremely unlikely that all the joints mentioned can be permanent and reliable, or even metal-to-metal because of the presence of, or likely insertion of plastics washers or plumbing parts and therefore one would expect to see supplementary equipotential bonding conductors between taps and sink body and between waste pipe and sink body. Should there be any similar doubt about the cold and hot water pipe systems then it becomes necessary to run a separate bonding conductor back to the main bonding conductor.

*Regulation 471-35* makes it quite clear, however, that even if the joints referred to would give the desired permanency and reliability, if the location concerned is a room containing a fixed bath or shower, supplementary bonding *must* be carried out. This requirement is still valid even if there are no exposed conductive parts present in that room.

Socket outlets are not permitted in a room containing a fixed bath or shower, as prescribed in *Regulation 471-34*, and it should be noted that *Regulation 471-36* requires a disconnection time within 0·4 s in the event of an earth fault associated with fixed equipment in such a room.

The purpose of this supplementary bonding is to reduce the voltages occurring between conductive parts should an earth fault occur not only in the circuit feeding the room containing the bath or shower but anywhere in the installation. This bonding therefore reduces the shock risk and is of particular importance in such rooms bearing in mind the reduced body resistance of persons using the bath or shower. This supplementary bonding must not be confused with the bonding required by *Regulation 525-10* between the metalwork of the electrical installation and exposed metal work of other services (*Regulation 525-10* admits the alternative solution of segregation). This bonding or segregation is intended to minimise the possibility of sparking occurring between the metalwork of the installation and exposed metalwork of the other services.

Particularly in industrial installations, there may be a great deal of fortuitous supplementary bonding but even where it is deliberately introduced it becomes difficult if not impossible to calculate the distribution of the fault current should an earth fault occur. Thus *Regulations 547-4* to *547-7* inclusive which deal with the determination of the cross-sectional area of supplementary bonding conductors do so on a quite arbitrary basis but it is pointed out that there may be instances where the minimum size of these conductors may need to be increased if a PME facility is used.

## First Edition of the Wiring Regulations

### Society of Telegraph Engineers and of Electricians

**RULES AND REGULATIONS**
**FOR THE PREVENTION OF FIRE RISKS ARISING FROM ELECTRIC LIGHTING,**

Recommended by the Council in accordance with the Report of the Committee appointed by them on May 11, 1882, to consider the subject.

**MEMBERS OF THE COMMITTEE.**

Professor W.G. Adams, F.R.S., *Vice-President.*
Sir Charles T. Bright
T. Russell Crampton
R.E. Crompton.
W. Crookes, F.R.S.
Warren De la Rue, D.C.L., F.R.S.
Professor G.C. Foster, F.R.S., *Past President.*
Edward Graves.
J.E.H. Gordon.
Dr. J. Hopkinson, F.R.S.
Professor D.E. Hughes, F.R.S., *Vice-President.*
W.H. Preece, F.R.S., *Past President,*
Alexander Siemens.
C.E. Spagnoletti, *Vice-President.*
James N. Shoolbred.
Augustus Stroh.
Sir William Thomson, F.R.S., *Past President.*
Lieut.-Colonel C.E. Webber, R.E., *President.*

These rules and regulations are drawn up not only for the guidance and instruction of those who have electric lighting apparatus installed on their premises, but for the reduction to a minimum of those risks of fire which are inherent to every system of artifical illumination.

The chief dangers of every new application of electricity arise mainly from ignorance and inexperience on the part of those who supply and fit up the requisite plant.

The difficulties that beset the electrical engineer are chiefly internal and invisible, and they can only be effectually guarded against by "testing," or probing with electric currents. They depend chiefly on leakage, undue resistance in the conductor, and bad joints, which lead to waste of energy and the production of heat. These defects can only be detected by measuring, by means of special apparatus, the currents that are either ordinarily or for the purpose of testing, passed through the circuit. Bare or exposed conductors should always be within visual inspection, since the accidental falling on to, or the thoughtless placing of other conducting bodies upon such conductors might lead to "short circuiting," or the sudden generation of heat due to a powerful current of electricity in conductors too small to carry it.

It cannot be too strongly urged that amongst the chief enemies to be guarded against, are the presence of moisture and the use of "earth" as part of the circuit. Moisture leads to loss of current and to the destruction of the conductor by electrolytic corrosion, and the injudicious use of "earth" as a part of the circuit tends to magnify every other source of difficulty and danger.

The chief element of safety is the employment of skilled and experienced electricians to supervise the work.

**I. THE DYNAMO MACHINE.**

1. The dynamo machine should be fixed in a dry place.
2. It should not be exposed to dust or flyings.
3. It should be kept perfectly clean and its bearings well oiled.
4. The insulation of its coils and conductors should be perfect.
5. It is better, when practicable, to fix it on an insulating bed.
6. All conductors in the Dynamo Room should be firmly supported, well insulated, conveniently arranged for inspection, and marked or numbered.

**II. THE WIRES.**

7. Every switch or commutator used for turning the current on or off should be constructed so that when it is moved and left to itself it cannot permit of a permanent arc or of heating, and its stand should be made of slate, stoneware, or some other incombustible substance.

8. There should be in connection with the main circuit a safety fuse constructed of easily fusible metal which would be melted if the current attain any undue magnitude, and would thus cause the circuit to be broken.

9. Every part of the circuit should be so determined, that the gauge of wire to be used is properly proportioned to the currents it will have to carry, and changes of circuit from a larger to a smaller conductor, should be sufficiently protected with suitable safety fuses so that no portion of the conductor should ever be allowed to attain a temperature exceeding 150°F.

N.B.–These fuses are of the very essence of safety. They should always be enclosed in incombustible cases. Even if wires become perceptibly warmed by the ordinary current, it is a proof that they are too small for the work they have to do, and that they ought to be replaced by larger wires.

10. Under ordinary circumstances complete metallic circuits should be used, and the employment of gas or water pipes as conductors for the purpose of completing the circuit, should in no case be allowed.

11. Where bare wire out of doors rests on insulating supports it should be coated with insulating material, such as india-rubber tape or tube, for at least two feet on each side of the support.

12. Bare wires passing over the tops of houses should never be less than seven feet clear of any part of the roof, and they should invariably be high enough, when crossing thoroughfares, to allow fire escapes to pass under them.

13. It is most essential that the joints should be electrically and mechanically perfect. One of the best joints is that shown in the annexed sketches. The joint is whipped around with small wire, and the whole mechanically united by solder.

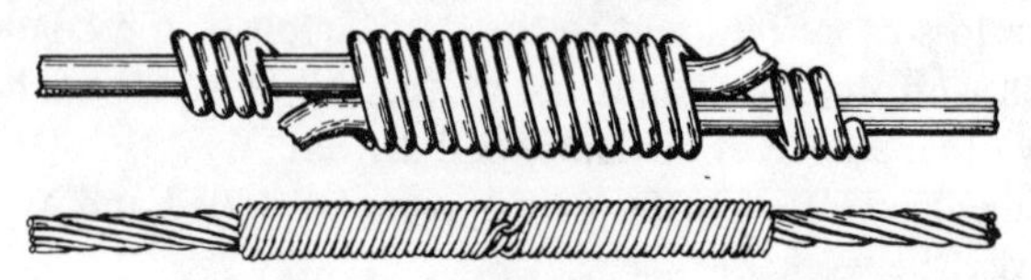

**Fig. 35** *Type of electrical joint recommended in the First Edition*

14. The position of wires when underground should be efficiently indicated, and they should be laid down so as to be easily inspected and repaired.

15. All wires used for indoor purposes should be efficiently insulated.

16. When these wires pass through roofs, floors, walls, or partitions, or where they cross or are liable to touch metallic masses, like iron girders or pipes, they should be thoroughly protected from abrasion with each other, or with the metallic masses, by suitable additional covering; and where they are liable to abrasion from any cause, or to the depredations of rats or mice, they should be efficiently encased in some hard material.

17. Where wires are put out of sight, as beneath flooring, they should be thoroughly protected from mechanical injury, and their position should be indicated.

N.B.–The value of frequently testing the wires cannot be too strongly urged. It is an operation, skill in which is easily acquired and applied. The escape of electricity cannot be detected by the sense of smell, as can gas, but it can be detected by apparatus far more certain and delicate. Leakage not only means waste, but in the presence of moisture it means destruction of the conductor and its insulating covering, by electric action.

## III. LAMPS.

18. Arc lamps should always be guarded by proper lanterns to prevent danger from falling incandescent pieces of carbon, and from ascending sparks. Their globes should be protected with wire netting.

19. The lanterns, and all parts which are to be handled, should be insulated from the circuit.

## IV. DANGER TO PERSON.

20. To secure persons from danger inside buildings, it is essential so to arrange the conductors and fittings, that no one can be exposed to the shocks of alternating currents exceeding 60 volts; and that there should never be a difference of potential of more than 200 volts between any two points in the same room.

21. If the difference of potential within any house exceeds 200 volts, whether

the source of electricity be external or internal, the house should be provided outside with a "switch," so arranged that the supply of electricity can be at once cut off.

By Order of the Council.
F.H. WEBB, *Secretary.*

Offices of the Society,
4, The Sanctuary, Westminster,
June 21, 1882.

# Appendix B

## Classification of degrees of protection provided by enclosures

The system of classification used in the Wiring Regulations to indicate the degree of protection provided by the enclosures of electrical equipment is that developed by the International Electrotechnical Commission and subsequently adopted by the British Standards Institution, being published as BS 5490. The IP Classification, as it is sometimes called, is being increasingly used in product standards and there is a close correlation between it and certain of the external influences classified for the international wiring rules and detailed in Appendix 6 of the Wiring Regulations.

In the IP Classification, the degree of protection is denoted by the letters 'IP' followed by two characteristic numbers the first indicating the degree of protection of persons against contact with, or approach to, live parts and against contact with moving parts (other than smooth rotating shafts and the like) inside the enclosure and protection of the equipment itself against ingress of solid foreign bodies.

The second characteristic number indicates the degree of protection for the equipment inside the enclosure against harmful ingress of water.

Thus a typical designation in this system would be 'IP 34'. When it is required to indicate a class of protection by only one characteristic numeral the omitted numeral is replaced by the letter 'X' e.g. IP 4X or IP 5X.

The significance of the letter 'X' is that although the equipment may have some degree of protection against the hazard concerned (in the two examples given above, the hazard is ingress of water) – but because of the intended location of the equipment the degree of protection does not require to be assessed. Either characteristic number may be zero so that one can have, for example, IP 20 or IP 40 and in these cases '0' means that there is known to be no protection against the ingress of water.

The following gives details of those classifications which appear in the Wiring Regulations.

*IP 2X*

The first numeral '2' indicates that fingers or similar objects exceeding 80mm in length and solid objects exceeding 12mm in diameter are excluded from the enclosure.

It also indicates that the enclosure has withstood a type test where the standard test finger shown in Fig. 36 (Test Finger I of BS 3042) was pushed with a force of less than 10N into openings in the enclosure, was placed in every possible position if it entered any opening, but adequate clearance was maintained between the test finger and live parts and moving parts inside the enclosure. The enclosure has also withstood a type test where a rigid sphere of $12 \ ^{+0 \cdot 05}_{-0}$ mm was applied to the openings of the enclosure with a force of 30 N ± 10% and did not pass through any of them. Again adequate clearance was maintained to live or moving parts inside the enclosure.

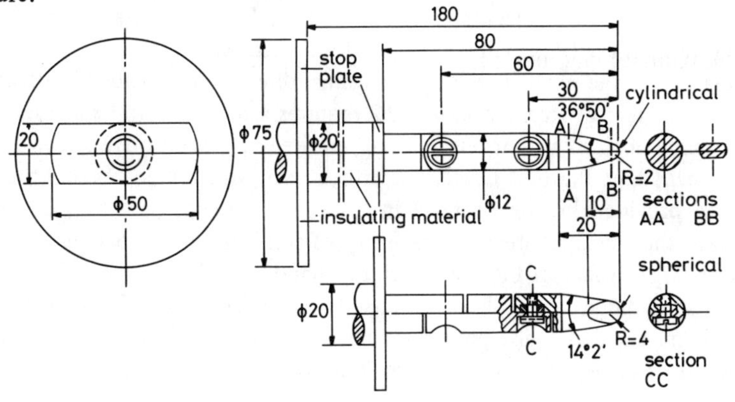


**Fig. 36** *Standard Test Finger 1 to BS 3042*
(By courtesy of British Standards Institution)
Dimensions are in millimetres
Tolerances: on angles ± 5′
on linear dimensions: less than 25 mm $^{+0}_{-0{\cdot}05}$
over 25 mm ±0·2

*IP 4X*

The first numeral '4' indicates that wires or strips of thickness greater than 1 mm and solid objects exceeding 1 mm in diameter are excluded from the enclosure.

It also indicates that the enclosure has withstood a type test where a straight stiff steel wire of 1·0 ± 0·03 mm diameter was applied with a force of 1 N ± 10% to all openings in the enclosure and did not enter.

*IP 5X*

The first numeral '5' indicates that the equipment concerned is 'dust-protected' and has withstood a dust test involving the use of talcum powder which after the test has not accumulated in a quantity or location such that it could interfere with the correct operation of the equipment.

Two categories of enclosure are recognised in this particular IP classification, the first being where the normal working cycle of the equipment concerned causes reductions in air pressure within the enclosure below the surrounding atmospheric pressure, e.g. thermal cycling effects, and the second where such pressure reductions do not occur.

Many product standards covering safety requirements published before the IP classification standard include tests using the standard test finger and a similar but rigid one in order to determine the non-accessibility of live parts inside enclosures and have additional tests using the test pins shown in BS 3042. Eventually these product standards may require to be changed to adopt the IP classification.

Similarly, many product standards, as regards protection against ingress of water, use the terms 'ordinary', 'drip proof', 'splash proof' and 'watertight'. Here again these product standards may be required to be changed to adopt the IP classification.

As regards protection against moving parts, BS 3042 indicates that Test Finger II in that standard, which is similar to Test Finger I but has a shorter distance to a single guard plate, is not acceptable to HM Factory Inspectorate and for this reason a further Test Finger IV is included in the British Standard.

In relation to the protective measure against direct contact called 'protection by placing out of reach', *Regulation 412-12* requires arm's reach to be measured from an intervening obstacle if that obstacle has a degree of protection less than IP 2X.

This means that the first numeral is either zero, which indicates that there is no protection provided by the obstacle in respect of persons or it may be unity, in which case the obstacle prevents the entry of solid objects exceeding 50 mm diameter or a large surface of the body such as a hand.

The first numeral could also be omitted and replaced by 'X' which, as already explained, means that while there may be some degree of protection of persons given by the obstacle it has not been assessed.

# Appendix C

## Voltages under earth fault conditions

This Appendix deals in general terms with the voltages appearing on conductive parts when an earth fault occurs and some reference is made to the alternative method given in Appendix 7 of the Wiring Regulations for compliance with *Regulation 413-3*.

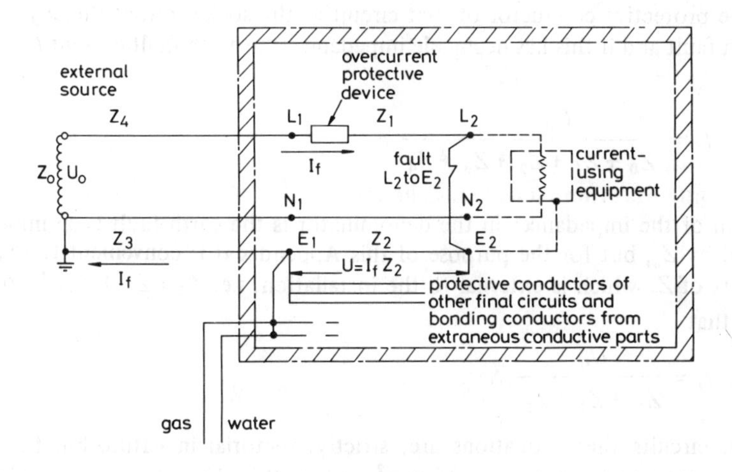


**Fig. 37** *Basic schematic diagram for TN-S system*

Fig. 37 is the basic or outline schematic diagram for a single-phase TN-S system where an external source of energy is supplying an installation within a particular location represented by the shaded outline.

Terminals $L_1$, $N_1$ and $E_1$ are the consumer's phase, neutral and earthing terminals, respectively, and those marked $L_2$, $N_2$ and $E_2$ are the corresponding terminals at a socket outlet or other point of utilisation. As shown, the exposed conductive parts of current-using equipment fed from the socket outlet and of the socket itself are connected by means of the protective conductor of the circuit to terminal $E_1$, as required by *Regulation 413-8*.

Similarly, the protective conductors of any other circuits in the location are also connected to terminal $E_1$ and this terminal, in its turn, is connected by means of a protective conductor (which may be the metallic sheath and armouring of the distribution supply cable) to the earthed point of the source.

The pipework of other services entering the location, such as the water and gas services, together with other extraneous conductive parts, is bonded to the consumer's earthing terminal $E_1$, as required by *Regulation 413-2*, thus converting the location into an equipotential zone.

Let

$U_0$ = nominal voltage to earth, in volts (for the source illustrated in Fig 37, $U_0$ is also the nominal phase-to-neutral voltage)
$Z_0$ = internal impedance of source, Ω
$Z_1$ = impedance of phase conductor of final circuit, Ω
$Z_2$ = impedance of protective conductor of final circuit, Ω
$Z_3$ = impedance of protective conductor of distribution cable, Ω
$Z_4$ = impedance of phase conductor of distribution cable, Ω
$I_f$ = earth fault current, A

As shown in Fig. 37, the phase conductor of the final circuit has come into contact with the protective conductor of that circuit at the socket outlet thereby creating an earth fault and if this has negligible impedance, the earth fault current $I_f$, is given by

$$I_f = \frac{U_0}{Z_0 + Z_1 + Z_2 + Z_3 + Z_4} \text{ A}$$

The sum of the impedances in the denominator is the earth fault loop impedance denoted by $Z_s$, but for the purpose of this Appendix it is convenient to segregate that part of $Z_s$ which is external to the installation, i.e. $Z_0 + Z_3 + Z_4$, denoted by $Z_E$, so that

$$I_f = \frac{U_0}{Z_E + Z_1 + Z_2} \text{ A}$$

For a.c. circuits these equations are, strictly, vectorial in nature but for cables having a cross-sectional area of 35 $mm^2$ or less, their inductances can be ignored and their resistances ($R_1$ and $R_2$ for the phase and protective conductors, respectively) used instead of their impedances in determining the earth fault current.

In order to be able to predetermine, for a particular installation, what the earth fault currents may be in the various final circuits, it is essential that $Z_E$ is either known, or at least some indication should be obtained at the design stage of the installation of the possible magnitude of this impedance.

Further, $Z_E$ should ideally be divided into its resistive and reactive components, but this is not possible where the supply is obtained from the public supply network. In any event a precise calculation of $Z_E$ is not possible due to the existence of fortuitous earth paths external to the installation such as those via metallic water pipes and gas pipes.

Thus the minimum value of earth fault current is given by

$$I_{f(min)} = \frac{U_0}{Z_E + R_1 + R_2} \text{ A}$$

where the denominator is the arithmetic sum of $Z_E$, $R_1$ and $R_2$, i.e. making the assumption that $Z_E$ is wholly resistive.

The maximum value of this current is given by

$$I_{f(max)} = \frac{U_0}{\sqrt{Z_E^2 + (R_1 + R_2)^2}} \text{ A}$$

i.e. making the assumption that $Z_E$ is wholly reactive.

While the earth fault is allowed to persist, it will be seen from Fig. 37 that a voltage, $U$ volts, will exist between the terminal $E_2$ and $E_1$ and it is this voltage to which the person protected would be subjected if he touched terminal $E_2$ or any exposed conductive part connected to it at the same time as he touched any conductive part (either exposed or extraneous) connected to the main earthing terminal $E_1$.

The current-using equipment shown by the dotted outline in Fig. 37 could be perfectly sound (and not even switched on) but while the earth fault at the socket outlet remained, a person touching that equipment and, for instance, a central heating radiator, would be subjected to that voltage $U$ volts. It can be readily shown that voltages can also exist between the exposed conductive parts of two sound equipments fed from a circuit if an earth fault occurs in that circuit or in any item of equipment fed by the circuit.

As shown in Fig. 37, the voltage appearing between the earth terminal of the socket outlet or the exposed conductive parts of the equipment connected to that terminal and exposed or extraneous conductive parts connected directly to the main earthing terminal is given by

$$U = I_f R_2 \text{ V}$$

This will have a minimum value, adding the components of the earth fault loop impedance arithmetically, of

$$U_{(min)} = \frac{U_0 R_2}{Z_E + R_1 + R_2} \text{ V}$$

and a maximum value

$$U_{(max)} = \frac{U_0 R_2}{\sqrt{Z_E^2 + (R_1 + R_2)^2}} \text{ V}$$

When the protective conductor and its associated phase conductor are of the same material and are run together over the same route, the ratio $R_2/R_1$ is a constant, denoted by '$m$', and it can be shown that the voltage $U_{(max)}$ cannot exceed $m\,U_0/(m + 1)$ volts.

For p.v.c.-insulated cables to BS 6004, '$m$' varies between unity and 2·67, and for the smaller sizes of mineral-insulated cables '$m$' is less than unity. Where the armouring of p.v.c.-insulated cables to BS 6346 is used as the protective conductor '$m$' may be as high as 5 in the range of cross-sectional areas up to 35 mm$^2$, and for the larger sizes '$m$' becomes even greater.

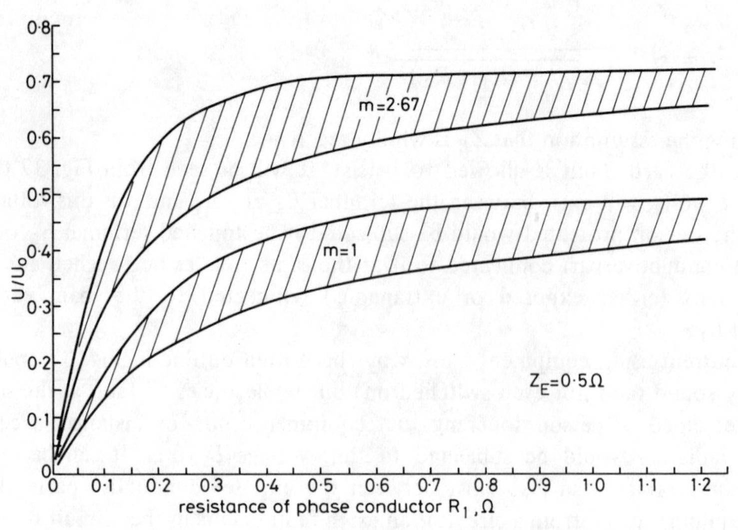


**Fig. 38** *Illustrating variation of $U/U_0$ with resistance of phase conductor*

Using the foregoing equations Fig. 38 has been developed for $Z_E = 0{\cdot}5\ \Omega$ and shows the envelopes of the ratio $U/U_0$ when $m = 1$ and when $m = 2{\cdot}67$. The upper limit of the former is asymptopic to $U/U_0 = 0{\cdot}5$ and that of the latter is asymptopic to $U/U_0 = 0{\cdot}727$.

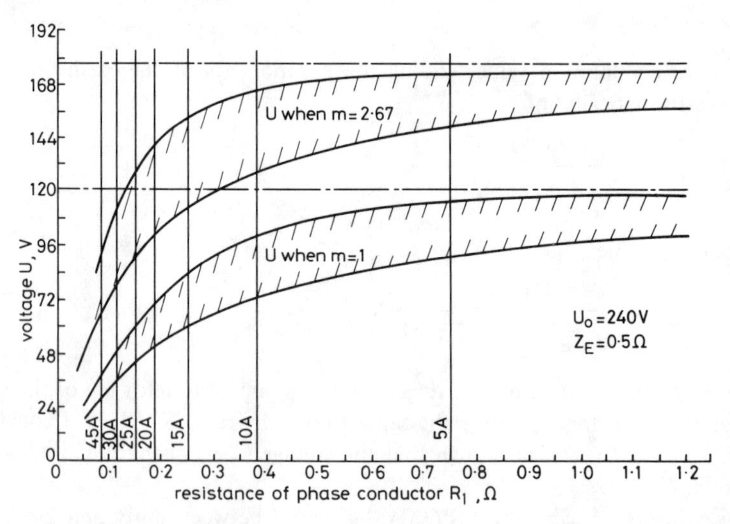


**Fig. 39** *Illustrating variation of U with resistance of phase conductor, when $U_0$ = 240 V*

Fig. 39 repeats Fig. 38 but for the specific case when $U_0$ = 240V, and it also indicates the maximum values of the resistances of the phase conductor for compliance with the limitation of voltage drop under normal load conditions, as prescribed in *Regulation 522-8*, assuming that the design current of the circuit is equal to 80% of the rated current of the overcurrent protective device concerned and that the circuit is single-phase.

Should $Z_E$ be less than the value (0·5 Ω) chosen as the basis for Figs.. 38 and 39, although the upper limits of the two envelopes remain unchanged, the initial slopes of those envelopes become steeper. For instance, if $Z_E$ = 0·2 Ω, the ratio $U/U_0$ for the case where $R_1$ = 0·10 Ω increases to approximately 0·64.

For circuits feeding a number of socket outlets the maximum voltage appearing under earth fault conditions would occur when the fault was at the socket outlet furthest from the origin of the circuit and would be between the exposed conductive parts of that socket outlet (or of current-using equipment fed from it) and extraneous conductive parts.

All the foregoing equations assume that the person protected remains with the equipotential zone and is not directly in contact with the general mass of Earth but even so, the voltages appearing between exposed conductive and extraneous conductive parts in the event of an earth fault (and not only those conductive parts associated with the faulty equipment) can be of such a magnitude to constitute a serious shock risk. Overcurrent protective devices will give adequate protection against that risk provided that the earth fault loop impedance is sufficiently low to give rapid disconnection of the supply in the event of an earth fault and for socket outlet circuits *Regulation 413-4* requires that the disconnection time shall not exceed 0·4 s.

The Wiring Regulations, as already indicated in this Commentary, in alignment with the international wiring rules, differentiate between socket outlet circuits and those supplying fixed equipment and for the latter require a maximum disconnection time of 5 s. The reason for this is that, in the former type of circuit, a socket outlet may well be used to supply equipment requiring to be gripped continuously in operation such as a portable drill or sander and this aspect represents a much greater risk to the user than is the case with fixed equipment.

The severity of any electric shock depends on the value of current passing through the body of the person experiencing that shock and its duration. Although the relationship between shock current and voltage is not linear because body resistance itself varies with voltage, the time for which an earth fault may be allowed to persist decreases as the voltage appearing between simultaneously accessible conductive parts increases.

For a single-phase socket outlet circuit when $U_0$ = 240 V, if the value of $Z_E$ is of the order of 0·8 Ω or less and the circuit complies with the limitation of voltage drop under normal load conditions, the maximum disconnection time of 0·4 s gives a reasonably practicable method of application of the international concepts concerning disconnection times and voltages under earth fault conditions.

Those rules also include the provision that provided the person protected remained within the equipotential zone and under normal dry conditions, from the consideration of shock risk only, a voltage between simultaneously accessible parts of less than 50 V could be allowed to persist indefinitely, and if that voltage is 50 V the disconnection time must not exceed 5 s (for socket outlet circuits).

It is these provisions which form the basis of the alternative method detailed in Appendix 7 of the Wiring Regulations which limits the impedance of the protective conductor of the circuit concerned.

It will be recalled that the limiting values of earth fault loop impedance to give a disconnection time of 0·4 s assume that the earth fault itself is of negligible impedance. They also assume that the position of the earth fault is such that no part of the normal load impedance is introduced into the earth fault loop. Should the earth fault possess some impedance or if there is part of the normal load impedance introduced into the earth fault loop, while the voltages appearing between simultaneously accessible parts decrease compared with the calculated values, so will the earth fault current leading to increased disconnection times, but such increased times using the alternative method do not lead to an unacceptable lowering of the safety level so far as protection against indirect contact is concerned. The alternative method automatically ensures that if, for any reason, the disconnection time exceeds 5 s, the voltage appearing between simultaneously accessible parts in the event of an earth fault will be less than 50 V.

Notwithstanding this, it was felt that it would be undesirable for an earth fault to be allowed to persist indefinitely, for instance because of the need to meet the thermal constraints in Chapter 54 of the Wiring Regulations, and for this reason the alternative method also requires the earth fault loop impedance to be limited to such a value that the disconnection time does not exceed 5 s, but admittedly then assuming negligible fault impedance for one's calculations.

It may be thought that adopting the alternative method with its limitation of impedance of the protective conductor of the circuit concerned would lead to either having to use an increased cross-sectional area of that protective conductor or a shorter length of circuit.

This, however, is not necessarily so. For instance, for a single-phase 240 V socket outlet circuit it will be generally found where the rated current of the overcurrent protective device is 30A or less, $Z_E$ is of the order of 0·5 Ω, the '*m*' value does not exceed 2·67 and the circuit complies with the voltage drop limitation of 2·5% of the nominal voltage under normal load conditions, compliance with the 0·4 s disconnection time specified in *Regulation 413-4* by limitation of earth fault loop impedance to the appropriate value specified in Table 41A1 in *Regulation 413-5* means that the circuit also complies with the limiting value of protective conductor impedance determined for the alternative method.

The general equations for determining the limiting value of the resistance of the protective conductor $R_2$ to the main earthing terminal are

(*a*) for fuses, other than semi-enclosed fuses to BS 3036

$$R_2 \leqslant \frac{50}{\text{current for 5 s disconnection}} \ \Omega$$

irrespective of the magnitude of $U_0$ the nominal voltage to earth.

(*b*) for semi-enclosed fuses to BS 3036

$$R_2 \leqslant \frac{240}{\text{current for 0·04 s disconnection}}\ \Omega$$

where $U_0 \leqslant 240\text{V}$.

(*c*) for Type 1 miniature circuit breakers

$$R_2 \leqslant \frac{50}{4\,I_n}\ \Omega$$

where $I_n$ is the rated current of the miniature circuit breaker. This equation applies irrespective of the magnitude of $U_0$.

(*d*) for Type 2 miniature circuit breakers

$$R_2 \leqslant \frac{50}{7\,I_n}\ \Omega$$

where $I_n$ is the rated current of the miniature circuit breaker. This equation applies irrespective of the magnitude of $U_0$.

(*e*) for Type 3 miniature circuit breakers

$$R_2 \leqslant \frac{50}{9{\cdot}4\,I_n}\ \Omega$$

where $I_n$ is the rated current of the miniature circuit breaker. This equation applies irrespective of the magnitude of $U_0$.

All the above equations are for the case where the protective conductor is not in the same cable as the associated phase conductor(s) and the temperature of the protective conductor at the commencement of the fault is assumed to be 30°C. As explained elsewhere in this Commentary, when the protective conductor *is* in the same cable is the associated phase conductor(s) the temperature of the protective conductor at the commencement of the fault is assumed to be the maximum permissible operating temperature. In the latter case, the limiting values for $R_2$ in the above equations are to be multiplied by 0·85. This factor is, however, only strictly correct where the cable insulation is p.v.c.

The alternative method is primarily of use where the installation comprises only one equipotential zone, such as the typical domestic installation. When the installation comprises more than one such zone and there is a possibility of equipment fed from one zone being taken into another, in order to retain the advantage of the alternative method of limiting the voltage appearing between simultaneously accessible parts in the event of an earth fault (should the disconnection time exceed 5 s), it becomes necessary to associate the limitation of the protective conductor impedance with the point of common connection of the protective conductors of the zones concerned. This is illustrated in Fig. 40, which shows an installation having two equipotential zones, e.g. in two separate buildings or in two separate locations in the same building.

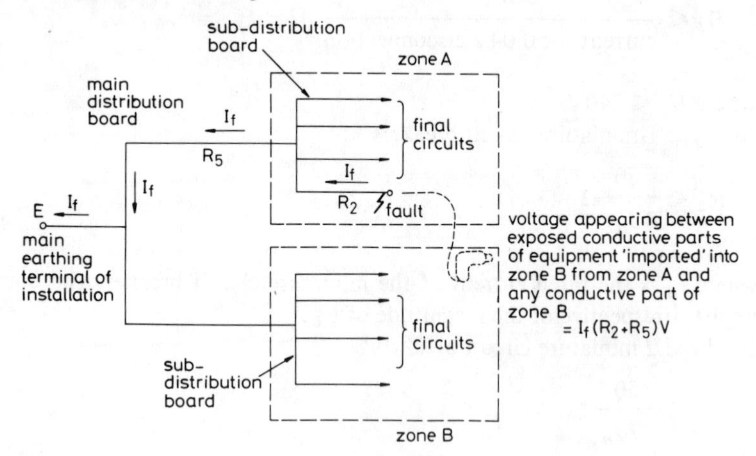


**Fig. 40** *Importing equipment from one equipotential zone into another zone*

If an equipment fed from a circuit in Zone A is taken into Zone B, and an earth fault occurs in that equipment or in the circuit feeding it, the voltage to which the person protected would be subjected would be $I_f(R_2 + R_5)$ V and not $I_fR_2$ V where, as before, $R_2$ is the resistance of the protective conductor of the circuit within the zone, i.e. to the main earthing terminal of the zone, and $R_5$ is the resistance of the protective conductor from that earthing terminal to the point of connection of that conductor with the protective conductor from Zone B, i.e. the main earthing terminal in the main distribution board.

This analysis has presupposed that there is no gas or water metallic pipework entering the two locations. If it did, then it would be required to be bonded to the main earthing terminal in each of the locations and in effect would act as supplementary bonding and the voltage to which the person protected would be subjected would reduce to $I_fR_2$ V.

Strictly, whether one uses the limitation of earth fault loop impedance method or the limitation of protective conductor impedance method for compliance with *Regulation 413-4*, the resistance of the supply flexible cables and cords of the current-using equipment should be taken into account. Unless it is known that these supply flexible cables and cords are going to be very long, it is suggested that this need not be done bearing in mind that the tabulated limiting values have been based on worst case conditions.

All the forgoing comments have been related to the TN-S system shown in Fig. 37 but they are equally pertinent to an installation fed from a PME supply, i.e. the installation and source of energy comprise a TN-C-S system. Consider now the situation which arises if, for the installation shown in Fig. 37, the current-using equipment was taken outside the equipotential location into say, the garden where the user of that equipment has now to be assumed to be in direct contact with the general mass of Earth.

Thus, if the earth fault occurs at the terminals of the socket outlet as shown in Fig. 37, although the fault current will be exactly the same as before, the voltage appearing on the current-using equipment is no longer $I_f Z_2$ V but $I_f(Z_2 + Z_3)$ V where $Z_3$ is the impedance, in the case of the TN-S system, of the metallic sheath and armouring of the supply cable, or in the case of the TN-C-S system, of the combined neutral earth of that cable.

In effect, the simultaneously accessible parts to be considered are the exposed conductive parts of the current-using equipment and the extraneous conductive part represented by the general mass of Earth. The magnitude of the impedance $Z_3$ for the case of an installation fed from the public supply network, for all practical purposes, is not available except perhaps for the TN-C-S system but because of the new requirement demanding protection of the circuit by means of a residual current device (if the circuit has a socket outlet or outlets specifically intended to supply equipment outside the zone) there is no need to calculate the voltage appearing in the exposed conductive parts of the equipment concerned and one makes the assumption that it may approach $U_0$, the nominal voltage to earth of the system. As explained in Chapter 4 of this Commentary the earth fault current will be comparatively large and certainly sufficient in the case of a 30 mA residual current device to disconnect the circuit in not longer than 0·04 s.

When considering an installation which is part of a TT system, it becomes necessary to differentiate between one which is fed directly from the public supply network and one which, for instance, takes a supply at high voltage and has the necessary step-down transformer as an integral part of the installation itself.

Limiting consideration here to the former type of installation, the resistance of the earth electrode at the supply undertakings substation, $R_B$, is now part of the earth fault loop impedance $Z_s$, as seen from Fig. 41.

The earth fault current $I_f$ is now given by

$$I_f = \frac{U_0}{Z_1 + Z_2 + R_A + R_B + Z_0 + Z_4} \text{ A}$$

As before, when the circuit conductors are less than 35 mm², $Z_1$ and $Z_2$ can be replaced by $R_1$ and $R_2$, respectively.

When it can be assumed that the person protected remains inside the equipotential zone, the voltage $U$ to which he would be subjected should he touch terminal $E_2$ or any exposed conductive part connected to the main earthing terminal $E_1$ is given by exactly the same equation given earlier in this Appendix for an installation which is part of a TN system, i.e.

$$U = I_f Z_2 \text{ V}$$

However, the earth fault current is considerably lower than for installations in TN systems and in most cases will be too low to give disconnection of the overcurrent devices in the times specified in *Regulation 413-4*. In order to give protection against indirect contact in an installation which is part of a TT system it is therefore necessary to use a residual current device or a voltage operated earth leakage circuit breaker, the former being the preferred device.

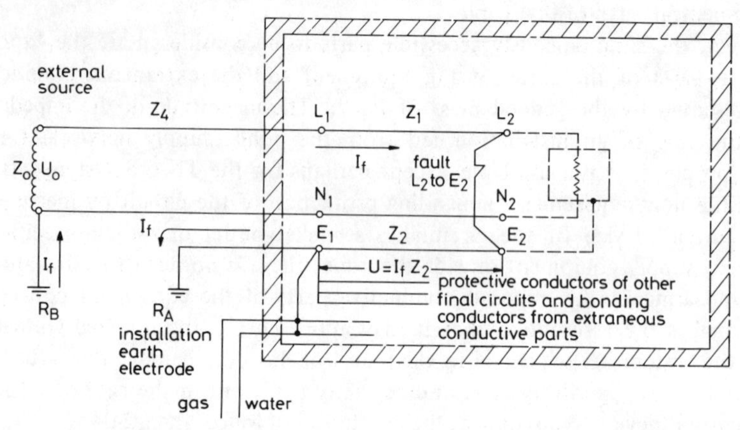


**Fig. 41** *Basic schematic diagram for TT system*

If the person protected moves outside the equipotential zone, the voltage to which he could be subjected in the event of an earth fault is $I_f(R_2 + R_A)$ V and as $R_A$ becomes a greater portion of the earth fault loop impedance $Z_s$ so will this voltage more closely approach the value $U_0$, and for installations which are part of a TT system the assumption is made that $U$ will, in fact, be equal to $U_0$.

## Ring final circuits

These notes are intended to assist the designer who wishes to use a ring circuit other than one of those detailed in Table 5A of Appendix 5 of the Wiring Regulations and they are primarily concerned with showing how to establish whether the ring circuit complies with the thermal requirements of *Regulation 543-2* and of *Regulation 434-6*.

Fig. 42 is the basic schematic diagram for the protective circuit of a ring final circuit supplying a number of socket outlets and in which an earth fault has occurred, as shown, at socket outlet A, some fraction '$x$' of the total route length of the circuit measured from the origin of the circuit. It will be seen that the circuit protective conductor presents two parallel paths for the earth fault current, $I_F$ A, which divides according to the usual rules for parallel circuits into two currents, $(1 - x) I_F$ A and $xI_F$ A.

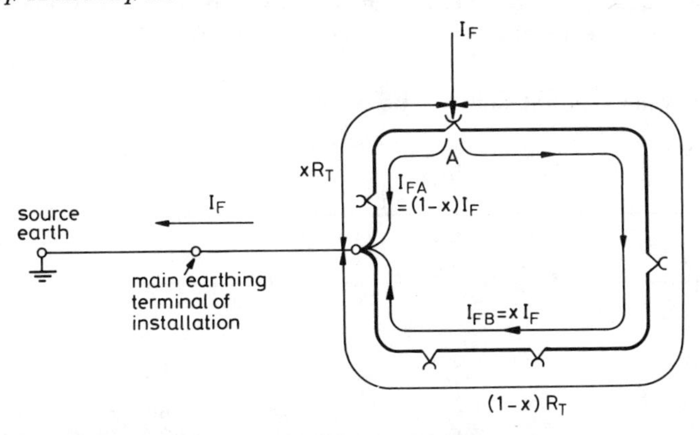


**Fig. 42** *Illustrating distribution of fault current in a ring circuit*

Assuming that the overcurrent protective device associated with the circuit is also intended to give protection against indirect contact so that in the event of an earth fault, the device must operate within 0·4 s, (as required by *Regulation 413-4(i)*), it is an easy matter to determine the limiting value of earth fault loop impedance, (if not already given in Table 41A1 of the Wiring Regulations) and then the total length of the ring circuit cable which could be tolerated. It is not quite so straightforward to determine whether the protective conductor of the ring circuit complies with the thermal limitations specified in *Regulation 543-2*, assuming that it is intended to use a protective conductor having a smaller cross-sectional area than its associated phase and neutral conductors, as would be the case if the ring circuit was wired in the so-called 'flat cables' to BS 6004.

The effective or equivalent resistance of the ring circuit protective conductor, $R_{eff}$ Ω is given by

$$R_{eff} = x(1 - x) R_T \ \Omega$$

where $R_T$ is the total impedance of the protective conductor, i.e. as measured between the ends of that conductor prior to them being connected together to complete the ring.

It can be readily shown that the maximum value of $R_{eff}$ occurs when $x = 0{\cdot}5$, i.e. corresponding to the earth fault occurring midway around the ring circuit so that it is then equal to $0{\cdot}25\ R_T\ \Omega$.

The associated phase conductor 'delivering' the current to the fault does so also through two parallel paths and therefore the effective resistance of that conductor follows the same pattern as the protective conductor. Assuming that the phase and protective conductors are of the same material and follow the same route the total resistance of the phase conductor is given by $R_T/m\ \Omega$ where '$m$' is the ratio of the cross-sectional areas of the phase and protective conductors.

For the flat cables to BS 6004 the '$m$' values are as shown in Table 24 which also gives, for convenience, the corresponding values of $(m + 1)/m$.

**Table 24: '*m*' values for flat cables to BS 6004**

| Cross-sectional area mm² | | | |
|---|---|---|---|
| Phase conductor | Protective conductor | $m$ | $\frac{m+1}{m}$ |
| 1 | 1 | 1 | 2 |
| 1.5 | 1 | 1.5 | 1.67 |
| 2.5 | 1.5 | 1.67 | 1.6 |
| 4 | 1.5 | 2.67 | 1.37 |
| 6 | 2.5 | 2.4 | 1.42 |
| 10 | 4 | 2.5 | 1.4 |
| 16 | 6 | 2.67 | 1.37 |

The earth fault loop impedance, $Z_s\ \Omega$, is the sum of the effective resistances of the phase and protective conductors plus the impedance $Z_E\ \Omega$ external to the circuit, i.e. on the supply side of the origin of the ring circuit, so that the general equation for $Z_s$ is

$$Z_s = Z_E + x(1-x)\frac{(m+1)}{m} R_T \ \Omega$$

However, when checking if the ring circuit meets *Regulation 413-4(i)*, it is only necessary to check if the minimum earth fault current is sufficient to cause operation of the overcurrent protective device.

Thus, if

$U_0$ = nominal voltage to earth, V
$I_F$ = earth fault current, A, to ensure disconnection in 0·4 s

Then

$$Z_E + 0{\cdot}25 \frac{(m+1)}{m} R_T \leqslant \frac{U_0}{I_F} \ \Omega$$

This equation shows how the total resistance $R_T$ of the ring circuit which can be tolerated is influenced by the value of $Z_E$, that part of the earth fault loop impedance on the supply side of the circuit.

In considering the thermal aspects of ring circuits under earth fault conditions, it will be found that there are three basic cases to be taken into account and these are now discussed but before so doing Fig. 43 shows a typical example of how the earth fault current $I_F$ A varies with the position of the earth fault relative to the origin of the ring circuit, i.e. with the value of '$x$'.

*Case 1*

This case is illustrated by Fig. 44 which shows the time/current characteristic for the fuse protecting the ring circuit, the 'adiabatic line' for the size and type of protective conductor it is intended to use and the curve of the variation of earth fault current with '$x$'. Here, as shown, the point of intersection of the time/current characteristic and the adiabatic line occurs at a disconnection time *longer than 0·4 s.*

This point of intersection gives, of course, the minimum earth fault current and hence the maximum earth fault loop impedance for the combination of the particular fuse and protective conductor concerned for compliance with the thermal requirements of *Regulation 543-2*. As already stated, the minimum earth fault current, which coincides with the point of maximum earth fault loop impedance must be of sufficient magnitude to ensure disconnection in 0·4 s and is therefore shown in Fig. 44 by $I_{F1}$.

Thus, in this case, irrespective of the location of the earth fault, a *single path* of the conductor size concerned would meet the adiabatic thermal requirement and the sole determining factors in designing the ring circuit are the necessity to comply with the maximum disconnection time of 0·4 s and with the voltage drop limitation in normal operation.

*Case 2*

In this case, illustrated in Fig. 45, the point of intersection of the time/current characteristic and the adiabatic line occurs in a time *less than 0·4 s*. As shown in Fig. 45, $I_{F1}$ is again the earth fault current for 0·4 s disconnection time, corresponding to the location of the fault being midway along the ring and $I_{F2}$ is the earth fault current corresponding to the point of intersection and when the location of the fault is the fraction $x_1$ from one end of the ring circuit or the fraction $x_2$ from the same end ($x_2$, of course, equals $(1 - x_1)$). It will be seen immediately that if the earth fault occurs in that part of the ring circuit up to the fractional $x_1$ from one end or beyond $x_2$ from the same end so that the earth fault current is equal to or greater than $I_{F2}$ then as for Case 1, a single path of the conductor size concerned would meet the adiabatic thermal requirement as prescribed in *Regulation 543-2*.

However, should the earth fault occur midway around the ring circuit such that $I_{FA} = I_{FB} = 0{\cdot}5\ I_{F1}$ A (see Fig. 42), then for the adiabatic thermal requirement to be met these must be less than $I_p$ A, as shown in Fig. 45.

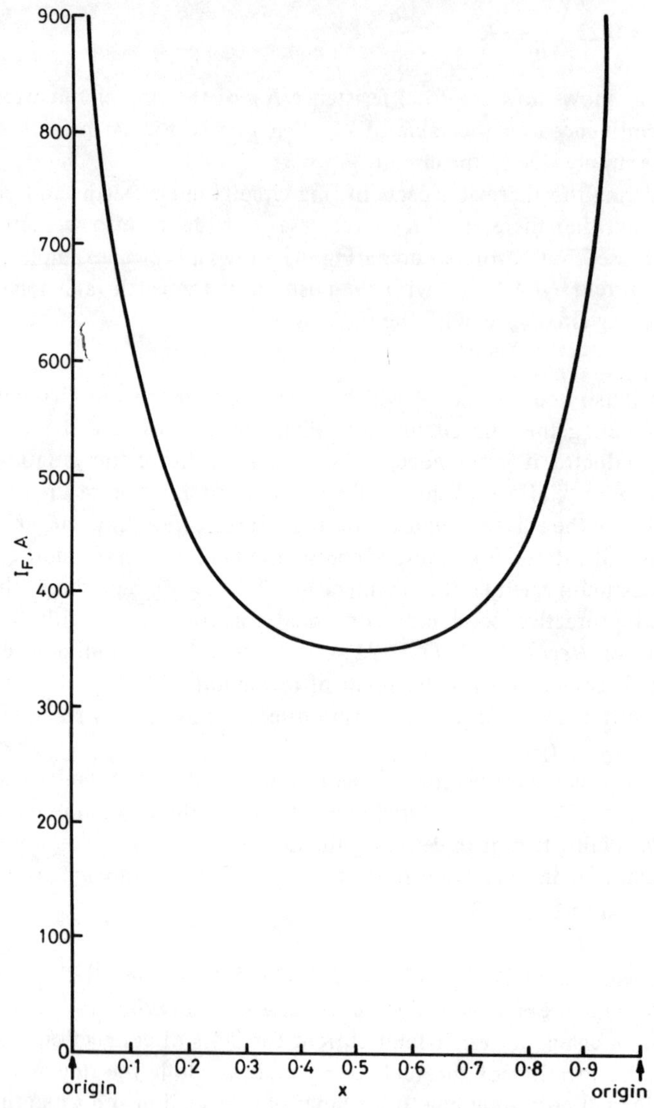


**Fig. 43** *Variation of fault current in a ring circuit depending on location of fault from origin of circuit*

$U_0 = 240$V

$Z_E = 0{\cdot}2\ \Omega$

$R_T = 2\Omega$

$I_F$ at origin of circuit $= \dfrac{240}{0{\cdot}2}$ A = 1200 A

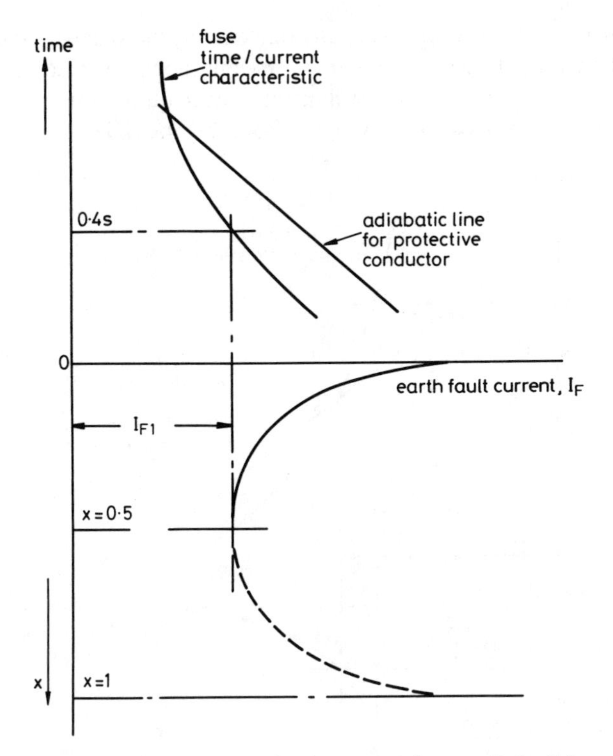


**Fig. 44** *Illustrating thermal requirements in the event of an earth fault in a ring circuit — Case 1*
Time and current axes are logarithmic; axis 'x' is linear

But from the adiabatic equation given in *Regulation 543-2*

$$I_p = \frac{SK}{\sqrt{t}} = \frac{SK}{\sqrt{0{\cdot}4}} \text{ A}$$

so that

$$0{\cdot}5\, I_{F1} \leqslant \frac{SK}{\sqrt{0{\cdot}4}} \text{ A}$$

from which

$$S \leqslant \frac{0{\cdot}316\, I_F}{K} \text{ mm}^2 \quad 2$$

In these equations

$S$ is the cross-sectional area of the protective conductor in mm$^2$

$K$ is the factor appropriate to the material of the protective conductor and its type (e.g. incorporated in the same cable as the associated phase conductor).

If the protective conductor cross-sectional area $S$ meets this equation, then if the earth fault did occur midway around the ring circuit, the adiabatic thermal requirement would be met but even if this is so, the question still remains as to whether the protective conductor would still meet that requirement if the earth fault occurred in the region between $x_1$ and $x_2$ (from the one end).

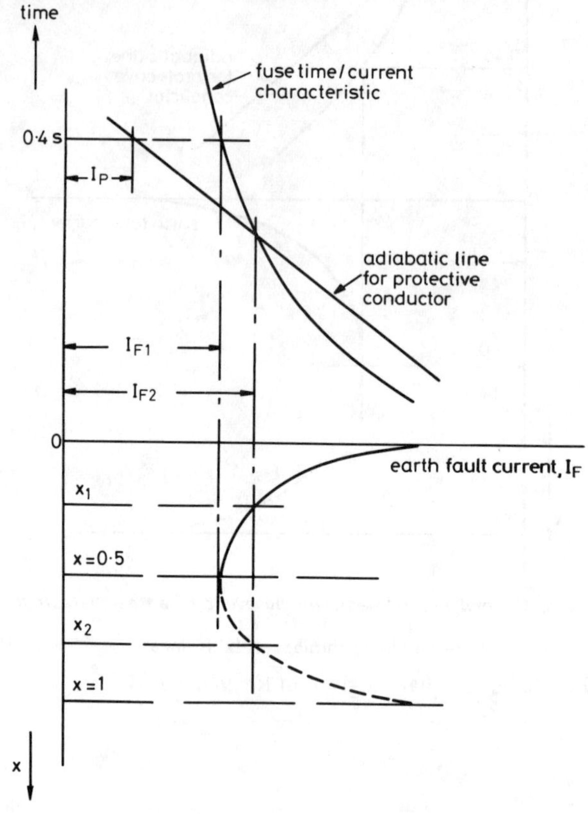


**Fig. 45** *Illustrating thermal requirements in the event of an earth fault in a ring circuit – Case 2*
Time and current axes are logarithmic; axis 'x' is linear

A rigorous proof is not necessary here but investigation of a few likely cases suggests that provided the cross-sectional area $S$ met the above equation, the thermal requirement would probably be met with the fault located in that region. If there is, in a particular case, any doubt, it is an easy matter to check by a graphical method if compliance would be attained.

Should the maximum earth fault loop impedance be such that the minimum fault current would be greater than $I_{F2}$ A, the situation would then be exactly as for Case 1 so that no matter where the earth fault occurred around the circuit, the adiabatic thermal requirements would be met.

*Case 3*

In this case there is no point of intersection between the time/current characteristic of the overcurrent protective device and the adiabatic line for the protective conductor cross-sectional area it is intended to use, the former being wholly above the latter, as shown in Fig. 46. For a radial circuit there would be complete failure to comply with the adiabatic thermal requirements but the question that has to be asked in relation to a ring circuit is whether or not the existence of the two parallel paths for the fault current could lead to compliance with those thermal requirements.

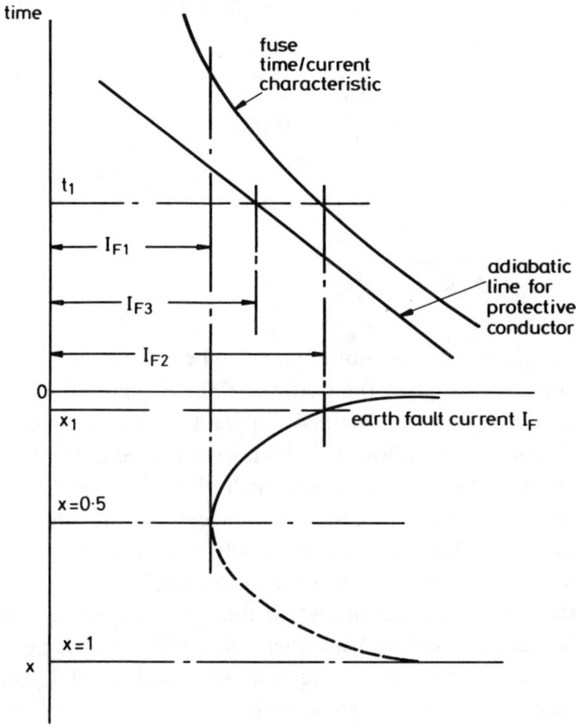


**Fig. 46** ***Illustrating thermal requirements in the event of an earth fault in a ring circuit—Case 3***
Time and current axes are logarithmic; axis 'x' is linear

From Fig. 46, if the earth fault occurs at some point $x$, the earth fault current is $I_{F2}$ A and this subdivides into its two parts, $x_1 \, I_{F2}$ A and $(1 - x_1) \, I_{F2}$ A, the latter being taken as the greater of the two (bearing in mind that when $x_1$ is greater than 0·5, the curve of fault current is merely the mirror image of that when $x_1$ is less than 0·5).

So that for compliance with the adiabatic thermal requirements

$$I_{F2} \, (1 - x_1) \leqslant I_{F3} \text{ A}$$

and therefore

$$I_{F2} \, (1 - x_1) \leqslant \frac{SK}{\sqrt{t_1}} \text{ A}$$

Examination of the time/current characteristics of semi-enclosed fuses given in Appendix 8 of the Wiring Regulations and repeated in Chapter 6 of this Commentary shows that for disconnection times of the order of 0·3 s and faster, the characteristics themselves can be expressed by the adiabatic equation, so that

$$I_{F2} = \frac{S_1 K_1}{\sqrt{t_1}} \text{ A}$$

Therefore

$$\frac{S_1 K_1}{\sqrt{t_1}} (1 - x_1) \leqslant \frac{SK}{\sqrt{t_1}} \text{ A}$$

from which

$$x_1 \geqslant \frac{S_1 K_1 - SK}{S_1 K_1}$$

The values of $S_1$ and $K_1$ are not required to be known, only their product, and the equation above readily gives the portion of the ring circuit in which the earth fault has to occur if the adiabatic thermal requirement is to be met. Or in other words, the last equation above indicates that where the time/current characteristic of the protective device does not intersect with the adiabatic line for the protective conductor it is intended to use, there will *always* be some portion of the ring circuit in which the earth fault can occur which will cause *non-compliance* with the adiabatic thermal requirements of *Regulation 543-2*.

One example of particular interest is that of a ring circuit using 2·5 mm$^2$ 'twin and earth' p.v.c.-insulated and sheathed cable which up to the recent amendment to BS 6004 had a 1 mm$^2$ protective conductor and which would normally be protected by means of a 30A semi-enclosed fuse.

From inspection of the time/current characteristic for that fuse when $t = 0{\cdot}1$ s, the earth fault current is 450 A so that the equation for this, the assumed adiabatic part of the characteristic, is

$$450 = \frac{S_1 K_1}{\sqrt{0{\cdot}1}}$$

that is

$$S_1 K_1 = 142$$

From Table 54C of the Wiring Regulations, assuming the cable conductors are of copper, the $K$ value for the protective conductor is 115.

So, compliance with *Regulation 543-2* is only obtained if the earth fault occurs at least

$$\frac{142 - 115}{142} . 100\% \text{ (i.e. } 19\%\text{) in from the origin of the circuit}$$

Thus 38% of that ring circuit can be said not to be protected from the thermal effects of an earth fault and it is for this reason that steps had to be taken in BS 6004 to increase the cross-sectional area of the protective conductor of the cable mentioned from 1 $mm^2$ to 1·5 $mm^2$ in order to retain the possibility of using a 30A semi-enclosed fuse to protect the circuit and still meet the thermal requirements of Chapter 54 of the Wiring Regulations.

# Appendix E

## The IT system

As shown in Fig. 3 in Chapter 3 of this Commentary, an IT system comprises a source of energy of which one point is either not earthed or is earthed through a high impedance and an installation in which the exposed conductive parts and extraneous conductive parts are earthed via a local earth electrode.

The supply undertakings in the United Kingdom directly earth their sources of energy, i.e. directly earth the neutral points of their distribution transformers, in order to meet the statutory requirement prescribed in Regulation 4 of the Electricity Supply Regulations 1937, or at additional points in the supply network if PME is used.

Thus no installation fed directly from the public supply network can, with the source of energy, comprise an IT system, and many electrical installation contractors and designers will never have any cause to use the regulations included in the Wiring Regulations concerning such a system.

IT systems, therefore, have a very limited, but nevertheless important, application, being used in industrial installations where continuity of supply is of particular importance, such as in continuous process plant.

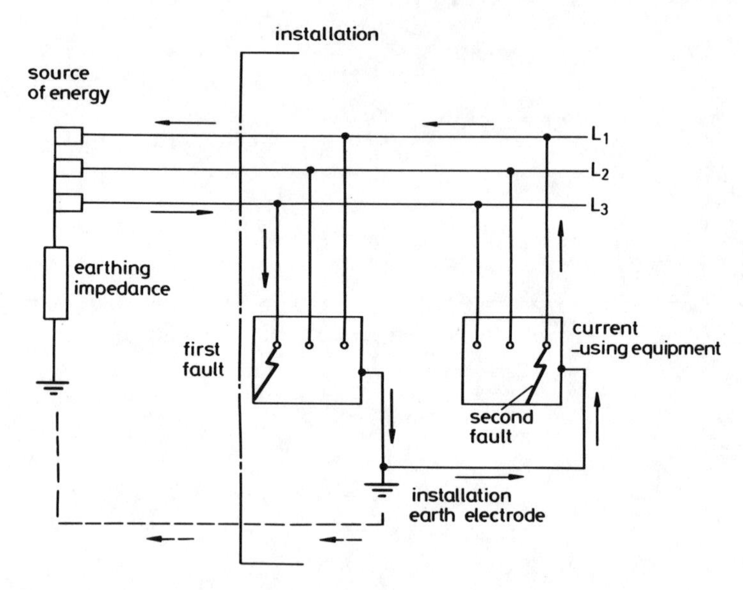


**Fig. 47** *IT system*
If an earthing impedance is used, the fault current shown by the broken arrow will be of the order of milliamperes and may be considered to be negligible compared with the current flowing through the first and second faults in the installation.

The suitability of the IT system for this application lies in the fact that, on the occurrence of an earth fault, either no fault current will flow if the neutral point of the source of energy is not earthed, or a low fault current which may be of the order of a few milliamperes will flow, if that neutral point is earthed through a high resistance. There is therefore no shock risk or thermal stress in the circuit conductors associated with this first earth fault and it is for this reason that *Regulation 413-16* requires an insulation monitoring device to be provided, allowing that device to either automatically disconnect the supply or to merely initiate an audible or visual signal.

This permits the installation supervisory staff to investigate and find the earth fault and, of course, hopefully to rectify it without having to disconnect the supply but as indicated in the note to *Regulation 413-16* the time for which the earth fault can be allowed to persist is limited and the fault should be eliminated as quickly as practicable.

The reason for this cautionary note is that while the earth fault is allowed to persist, the system is converted into an earthed system and as shown in Fig. 47, on the occurrence of a second earth fault, the system acts in the same way as a TN or TT system. If the source earth and the installation earth are connected together, the IT system acts as a TN system and the requirements for the latter type apply whereas if the source earth and the installation earth are not interconnected the requirements for a TT system apply, as prescribed in *Regulation 413-17*.

It is not intended to give here detailed analyses of the voltages which appear between simultaneously accessible parts and the fault current which flows, but these will depend on the constituent impedances of the fault loop and on whether the neutral is distributed or not. It is sufficient to comment that on the occurrence of the second earth fault there is the same order of shock risk and of thermal stress in circuit conductors as occurs in TN and TT systems and therefore it is essential that the protective devices giving protection against indirect contact (*Regulation 413-15* admits residual current devices and fault-voltage operated protective devices) give disconnection in the times specified in *Regulation 413-4*.

Of course, for an installation which is part of an IT system, main equipotential bonding complying with *Regulation 413-2* must be carried out, and particular attention should be paid to supplementary equipotential bonding.

## Appendix F

# Publications of the Electricity Council

The Electricity Council publishes Engineering Recommendations, Reports and Standards which are of considerable interest to engineers and others concerned with electrical installation practice and which are generally available.

### Engineering recommendations

*G5/3(1976):* Limits for harmonics in the UK electricity supply system.
*G12/1(1975):* National code of practice on the application of protective multiple earthing to low and medium voltage networks.
*G15 (1971):* Standardisation of electricity supply arrangements in industrialised component buildings.
*G16 (1972):* Standard specification for a new single-phase service termination/metering up to 100 A. Standard features of the assembly on a meter board (or equivalent) of cutout, meter(s), timeswitch, or other apparatus.
*G22/1 (1980):* Superimposed signals on Public Electricity Supply Networks. Sets down the ESI's attitude to devices using a consumer's electrical wiring installation and the public supply network as a communication or signalling channel.
*G26 (1975):* The installation and operational aspects of private generating plant.
Three main modes of operation are defined: (i) independent operation, (ii) parallel operation with the Electricity Board's system, and (iii) operation with alternative connection to the Electricity Board's system. Within these modes, the operation of plant, safety and technical requirements are reviewed, practices are recommended and typical protection arrangements given with supporting diagrams.
*P9 (1963):* Supply to welding plant. Amendment 1 (1975) Describes load characteristics of various types of welding plant and outlines practical methods of calculating the resulting voltage fluctuations at points on an electricity supply system.
*P13/1 (1979):* Electric motors – starting conditions.

### ACE reports

*ACE 7 (1963):* Supply to welding plant. Amendment 1 (1975) Amplification of Engineering Recommendation P9.
*ACE 15 (1970):* Harmonic distortion caused by convertor equipment.
*ACE 22 (1969):* Survey of the amplitudes and frequency of occurrence of

surges on m.v. systems.

*ACE 41 (1975):* Guide to an ERA code of practice for the avoidance of interference with electronic instrumentation and systems. Indicates the thinking behind the guidance given in those aspects of the Code of Practice relating to interference levels and susceptibility tests in respect of power supply disturbances.
*ACE 73 (1979):* Limits for Harmonics in the UK Electricity Supply systems.

**ESI standards**

*12-3:* Outdoor meter cupboards, Issue 1 1971
*43-30:* Low voltage overhead lines on wood poles, Issue 1 issued (1979)

Requests for all documents except ESI Standards should be made to the undermentioned address:

Head of Distribution Engineering Branch,
The Electricity Council,
30 Millbank,
LONDON SW1P 4RD.

Requests for ESI Standards should be made to the following address;

Engineering Document Unit,
Engineering Services Department,
Central Electricity Generating Board,
Courtenay House,
18 Warwick Lane,
LONDON EC4P 4EB.

# Index by Regulation Number

This index gives the page number(s) of the text in which reference is made to the Regulation(s) concerned. It does not include the Regulations included in the inspection list for metallic conduits in Chapter 11.

# Index